THE ACTIVITY OF YOUNG CHILDREN
DURING SLEEP: AN OBJECTIVE STUDY

UNIVERSITY OF MINNESOTA

THE INSTITUTE OF CHILD WELFARE

MONOGRAPH SERIES NO. XVIII

THE ACTIVITY OF YOUNG CHILDREN DURING SLEEP

An Objective Study

BY

CHESTER ROY GARVEY

LATE NATIONAL RESEARCH COUNCIL SCHOLAR
IN CHILD DEVELOPMENT

1939

THE UNIVERSITY OF MINNESOTA PRESS

MINNEAPOLIS

PRINTED IN THE UNITED STATES OF AMERICA

FOREWORD

This book is published posthumously. The author, a promising young scholar, died while preparing the results of this investigation for publication. The work of getting the monograph into its final form was done by members of the staff of the Institute of Child Welfare, with the able assistance of Professor Harry M. Johnson, formerly head of the Simmons Investigation of Sleep at the Mellon Institute and at present professor of psychology at Tulane University.

There still remain, however, certain points that would undoubtedly have been clarified by Dr. Garvey, but which cannot be changed without grave risk of misinterpreting the author's intended meaning. The study is therefore presented much as he left it. Every effort has been made to adjust whatever slight discrepancies the editors discovered and to put into simpler and more readable form material of a somewhat complex nature.

Nevertheless, the study as a whole is a valuable contribution in a field in which much work might still be done. Few previous studies of children's sleep are available in English, and none on the specific phases covered by this one. The author's method of calculating mean rest periods for children's night sleep, and the correlations he has worked out between activity during sleep and other variables, will undoubtedly prove extremely useful in later studies of similar phenomena. Also of great value are the excellent graphs, which tell the story of the author's findings most vividly and effectively.

Numerous "leads" for future investigations are pointed out all through this monograph, and the reader who is familiar with the literature in the field will doubtless discover many more that are not directly referred to. Any future worker on the sleep of young children will need to make a careful study of the material presented here, in regard both to methods and to findings.

JOHN E. ANDERSON
Director, Institute of Child Welfare
University of Minnesota

ACKNOWLEDGMENTS

The list of all who deserve credit for valuable assistance during the course of this study is entirely too long for publication. It gives me great pleasure, however, to name as many as space permits.

First of all, I wish to say a word in appreciation of the excellent manner and spirit in which the parents of my experimental subjects kept records of their children's sleep, uninterruptedly, four months. A study of this kind, which under ideal circumstances would be carried out in night nursery schools under controlled conditions, imposes much responsibility upon the parents.

The counsel of Professor Harry M. Johnson, head of the Simmons Investigation of Sleep, Mellon Institute, has been invaluable.

To Dr. John E. Anderson, director, and to the Institute of Child Welfare, University of Minnesota, I owe thanks for advice and assistance. To the Committee on Child Development of the National Research Council I express gratitude for the scholarships which made this work possible.

I wish to thank my many teachers and colleagues whose advice I have often sought, especially Dr. Florence L. Goodenough, who gave her time freely for statistical consultation; and Dr. Richard E. Scammon, who directed my study of graphics and of child anatomy. The late Dr. J. Arthur Harris' introduction to biometry has also been of great value in my work.

Thanks are due the Simmons Company and the Simmons Investigation of Sleep for the beds used in the study and for lending twelve recording instruments too costly to be obtained otherwise.

I am indebted also to Mrs. Ruby Glockler and Mr. Robert Challman for the use of data; to Mr. Robert Challman, Mr. Roger Brown Loucks, and Mr. Donovan Lawrence for methodological and technical advice and assistance; to Dr. Josephine Foster for permission to use the records of the nursery school; to my sister-in-law, Miss Lois M. Gillis of the University of Kansas, my sister, Miss Opal Garvey, R.N., and my wife, Helen Gillis Garvey, for assisting carefully and untiringly with the many statistical calculations and with the preparation of the manuscript. To my parents, Mr. and Mrs. W. W. Garvey, I am grateful for inspiration and assistance throughout the study.

C.R.G.

CONTENTS

fect of Weight and Body Build on Sleep—Effect of Posture
on Sleep—Effect of Ultra-Violet Rays on Sleep—Summary

THE ACTIVITY OF YOUNG CHILDREN
DURING SLEEP: AN OBJECTIVE STUDY

I. REVIEW OF THE LITERATURE

Sleep has been a topic of popular and scientific interest for many years. Anecdote and speculation on the subject comprise an enormous body of literature, and medical and scientific writings on sleep and the maladies of sleep have become too numerous in recent years to be mentioned in a study of this kind. Therefore only books and articles related to some of the particular problems encountered will be referred to here, in connection with the specific problems to which they apply. For reviews of the recent scientific literature one should see the articles by Johnson, Swan, and Weigand (27), Johnson and Swan (26), and Johnson (21) in the *Psychological Bulletin.*

Much of the scientific work, including the studies of Kleitman (34), Lee and Kleitman (39), Laslett (38), and Robinson (47, 48), has consisted in attempting to measure the physiological and psychological effects of loss of sleep. Other studies made during the process of sleep itself have involved sensory stimulation during sleep (4, 35, 40, 42, 49), measurement of changes in the electrical properties of the body (37), and recording of changes in bodily position (3, 14, 30, 36, 45, 55, 56, 58). The first studies of spontaneous activity seem to have been made with small animals by means of the squirrel-cage apparatus (53).

Still another method of study is to record the time of going to bed, the time of going to sleep, and the time of rising; from these to calculate the length of night, length of sleep, and length of initial restless period; and then to relate these to various factors, such as the age and sex of the subjects and the seasons of the year (8, 16). The present study employs both this method and that of registering changes in bodily position by instrumental means, but principally the latter.

Szymanski developed several variations of the kymographic recording type of apparatus, which he called *Aktographen,* for use with both animals and human beings (56, 57, 58, 59). It is true that one gets the impression from Herz (17) and Gerber (11, 12) that the kymographic type of apparatus got its start in the idea of a German engineer, Karl Nägele, based upon the principle of recording movements of the center of gravity. But actually Nägele's

3

machine seems to have been used first by Gerber (11) in 1921, whereas Szymanski had been working with his instruments since 1914 (57, 58). He used them to plot the distribution, over the twenty-four hours, of the activity periods and rest periods of small animals, insects, fish, and human infants (56, 58, 59). Gerber seems to have been the first to record activity during sleep alone, although Kreidl and Herz (36) made some observations of sleep periods alone on one of their subjects. Kreidl and Herz used Szymanski's setup to record the activity of the blind, but for the study of the deaf and deaf-blind they used Nägele's, which they regard as an improvement over Szymanski's because it imposes fewer artificial conditions upon the sleeper. Nägele's instrument, which is described by Gerber (11) with illustrations and diagrams, substitutes ink on white paper for the smoked drum.

Karger (30) used a modification of the Nägele apparatus in studying the sleep of children. The instruments of both Nägele and Karger registered the movements of the bedpost of a balanced bed. The movements were transmitted through a Marey tambour and a pneumatic tube to a kymograph. Szymanski used mechanical transmission of the movement on some of his apparatus for small animals, but for his work with infants he used the tambour.

H. M. Johnson developed several types of balanced mountings for beds, but finally decided upon a very simple method of direct mechanical transmission of the sleeper's movements to a strip of moving paper. (See Figures 1 and 2.) A recording pen is fastened to a small weight which hangs by a silk line from the top turn of one of the coil springs near the center of the bed spring. The weight and pen move up and down with the shifts in the weight of the subject upon the spring, thus making vertical shifts in the otherwise horizontal record on the moving paper. (Recording ink on white paper has been found much more satisfactory than smoked kymograph paper.) The apparatus has an advantage over the registering squirrel cage in that it gives a permanent record of the time at which each movement and the successive periods of activity occur.

Shirley (50) has maintained that these records cannot be used for a quantitative study of the total amount of activity. This is true, as Johnson and Swan (26) have also pointed out. It should be added, however, that the same criticism applies to the revolving drum apparatus, and for the same reasons. The revolving drum

FIGURE 1.—THE BED AND THE APPARATUS USED FOR MEASURING THE CHILDREN'S MOTILITY DURING SLEEP

FIGURE 2.—THE JOHNSON KINETOGRAPH

is subject to a period of its own, and it yields only to certain components of the animal's movements. The angular displacement of the cage is not, therefore, directly proportional to the energy expended by the animal moving in the cage. The records of Johnson's type of apparatus could be used for a quantitative study of the total amount of activity by integrating the displacement of the line acording to the same principle as that governing the cage, since the latter merely integrates certain angular displacements of the cage. We believe that Johnson's criticisms of this principle are valid, however, and that therefore it is better to confine ourselves to groups of movements, separated in time, which are large enough in themselves to be meaningful.

II. PURPOSES AND METHOD

PURPOSES

The main purposes of the present study were to set up norms (1) of the distribution of motility during sleep and (2) of the mean rest period during the sleep of young children in such a way that these norms can be used (a) as a basis for comparing the sleep of adults with that of children and (b) as a basis for comparing the effect of an experimental variable with normal "sleep activity." (This term indicates the activity that takes place during sleep. It characterizes the type of behavior dealt with in this study as contrasted with the subject matter of studies that deal with sleep per se.) An example of such a variable might be the occurrence of a high temperature, the administration of a drug, or the presence of noise or other form of stimulation. Findings pertaining to the first of these purposes are presented in the curves of Figures 3, 4, 5, and 8; those pertaining to the second, in Table 1 and in the Appendix. Our secondary purposes were to study the relationship between various factors and the intensity and distribution of nocturnal motility.

SUBJECTS AND CONDITIONS

The subjects were pupils in the nursery school of the Institute of Child Welfare of the University of Minnesota. They slept in their own homes. In all, 22 children were used, 8 girls and 14 boys, ranging in age from 25 to 58 months. Records were made for a year or more on 10 children (Subjects 3 to 12 inclusive), for 9 months on 1 (Subject 1), and for 5 months on 2 more (Subjects 2 and 22). These 13 children formed the main experimental group, upon which the curves of distribution of activity are plotted, and upon which are based most of the correlation studies of various factors. Nine additional children (Subjects 13, 14, 15, 16, 18, 19, 20, 21, and 32), who were used during periods ranging from 1 to 9 months, yielded from 25 to 74 records each. More detailed information concerning the subjects may be found in the Appendix.

The first record was made on the night of January 30, 1928, and the last on the night of January 21, 1930. The dates of the first

6

and the last record for each child, together with the child's birth date, are given in the Appendix. The data on Subject 32 were excluded from many of the tables and calculations because this subject suffered from a severe case of whooping cough throughout most of the period during which her sleep activity was being recorded. A total of 3,486 usable records were obtained, 2,188 on boys and 1,298 on girls.

Beds and bedding.—The beds in which the subjects slept were identical Simmons cribs. No difficulty was encountered in getting the children to accept the new beds. They liked them very much, except that one boy insisted that he was too large for his bed. It was hard to determine whether he really needed more room or merely objected to using a bed intended for babies and young children. At any rate his sleep was not found to be noticeably different from that of the other children.

The spring suspension on these beds is a combination link and coil construction. A wire chain link net is stretched between short coil springs attached to all four sides of the frame. The mattress, made to specification by the Simmons Company, differs from that company's regular mattress only in being bound in two separate parts. The lower part is a vertical coil spring mattress about five inches deep, while the upper part is a hair mattress pad about two inches thick. The mattress was covered with a rubber sheet, a thin blanket, and a bed sheet. The selection of the rest of the bedding was left to the discretion of the parents. The bedclothes were tucked under only the upper mattress so that the process of bedmaking would not disturb the recording mechanism, which was attached to the coil spring mattress. The instrument is noiseless and the child was unaware that it was attached to the bed. (See Figure 1.)

Daily reports.—The mother of each child was asked to start the recording instrument by turning a button each evening, to stop it the same way in the morning, and to make a daily report on blanks similar to the sample included here. The back of this card was used for recording various statistical data, including the actual distribution of active intervals among the total number of intervals, which was transferred directly from the machine record.

METHOD

In this study the method of Johnson (24; 26, pages 18–21) was used in all essential details in order that our findings on children could validly be compared with his on adults, or with those of any-one else using the same instruments and procedure. The recording instruments are the kinetographs designed by Johnson and built in his laboratory at the Mellon Institute. (See Figure 2.) In addition to the pen that records the movements of the child, these is a

SLEEP RECORD

INSTITUTE OF CHILD WELFARE

Date ....

Current on at .... P.M., off at .... A.M.

Child in bed at .... P.M., arose at .... A.M.

Temperature of room .... degrees at .... P.M.

.... degrees at .... A.M.

Number of windows open .... Distance open .... inches

Bedclothes: .... spread .... quilts

.... wool blankets .... cotton blankets

Bed wet ....

Other occupants of room ....

Health of child: well, fair, ill

Illness .... temperature ....

Bowels: loose, normal, constipated

Activities between supper and bed:

Violent exercise, quiet play, exciting stories

Nature of story ....

Emotional upset ....

....

Other activities ....

....

pen that writes simultaneously on the same paper and makes a small mark every five minutes, being actuated by an electromagnet that is thrown in circuit momentarily at the beginning of each five-minute interval by the works of a clock. Before the data were collected, it was determined that a displacement of the recording pen to the extent of one millimeter was the smallest displacement that could be accepted as evidence of a change in bodily position as great as the movement of a limb. Since we do not wish to take account of any stir of lesser magnitude, we have disregarded any shift in the record line of less than a millimeter. Whenever any shift of a millimeter or more occurs within a five-minute interval, that interval is classified as "active"; otherwise it is "passive." In the case of the active intervals, the movement or movements, whether one or many, are assumed to occupy all the first half and only the first half of the interval. This is the most reasonable assumption (the activity could be assumed to occupy all or none of the time within the interval), and on the average it is true.

In general, the records were treated in two ways. First, the temporal distribution of the active intervals, i. e., the distribution of motility over the time in bed, was plotted. Second, the mean length of time between successive groups of movements, i. e., the mean rest period, was calculated. For the distribution of motility, records of one hundred nights were assembled for each subject in the manner suggested by Johnson (26, page 20). The five-minute intervals on each record were numbered in order, beginning with the one during which the child was put to bed. The number of nights in which each of these intervals was found to be active was plotted against the number of that interval. When the series of interval numbers is translated into a scale of hours after time of retiring, the result is a curve of distribution of motility over the time in bed; samples are given in Figure 3. Arbitrarily dating the beginning of sleep from the beginning of the first period of inactivity greater than five minutes, we renumbered the five-minute intervals on each record beginning with the first passive interval, plotted against this number the number of nights in which each of these intervals was active, and thus obtained a curve of distribution of motility over the time asleep. Samples of these curves are shown in Figure 4.

For the mean rest period, each active interval is given a number which indicates its distance from the last preceding active interval in units of five-minute intervals. (The first interval in the night is always given the number 1 on the assumption that the child has moved within the preceding five minutes.) The sum of these interval numbers, then, is the total number of intervals and can be used as a check on the counting of the latter. This total is multiplied by five, the result is divided by the number of active intervals, and 2.5

minutes is subtracted from the quotient. This gives the mean rest period in minutes, or the average length of time from the middle of one active interval to the beginning of the next. The mean and standard deviation of the mean rest periods for each child for each month are recorded in the Appendix and are summarized for all subjects in Table 1 (page 14).

MEAN REST PERIODS

Mean rest periods are calculated in two ways. According to the first method, the length of night used (called T_1) is equal to the number of five-minute intervals between the time the child gets into bed at night and the time he leaves it in the morning. From this is obtained the mean rest period for all the time the child is in bed (called mean rest period 1, or simply MRP1). According to the second method, the length of night used (called T_2) is the number of five-minute intervals left after the initial and final periods of activity are deducted from the length used in the first method. The initial active period includes the number of intervals after the child goes to bed during each of which at least one movement occurs. Thus the record starts after the first rest period greater than five minutes. The final active period, if any, is that period just before arising, during which the child is continuously active (without any period of rest greater than 2.5 minutes), for more than 25 minutes. This means that if the child is active for more than 25 minutes just before arising, the record is terminated at the beginning of this continuous activity. This second mean rest period (called MRP2) is more likely to represent the length of time the child is asleep, since it is very unlikely that he is still awake when he becomes quiet for more than five minutes, and it is quite likely that he *is* awake during the time when he is very active just before getting out of bed.

This second criterion for terminating the record was decided upon because the point of termination indicated coincides with the point on the record at which the form of the tracing changes abruptly. Preceding this point, activity is so infrequent that its distribution can be measured in five-minute units. Beyond this point, activity is virtually continuous. The 25-minute provision saves a great deal of the extra labor involved in applying the above criterion, and since most of the final active periods that are less than 25 minutes are very short indeed, the error of ignoring them is not considered important. An error of even 25 minutes is less than 4 per cent of a typical night of 11 hours, and an error of this size is rare.

The abrupt change in the character of the tracing before the child arises is so obvious that it can be pointed out even by inexperienced observers. May it not indeed be a compound of a curve of sleeping and one of waking activity? If so, then the point of change in the curve would correspond to the point at which the child awakes. Fortunately, the mother of one of the children (Subject 3) made some notes that enable us to check the validity of our criterion of awaking. The mother's report for the night of April 12, 1928, carries a note that reads, "Awake 45 minutes before getting up." Our criterion ends the record 10 intervals short of the time the child got up, or about 47 minutes in this case. The report for May 4, 1929, says, "Awake between 5 and 6, up 6:20. Awake fully ¾ hr. before getting up." The criterion ends this record 9 intervals (about 46 minutes) before the time of arising, or about 5:34. Thus the only notes we have coincide very closely with our criterion.

As was pointed out, the use of five-minute intervals as units of measurement involves the assumption that any such interval that contains movement can be treated as if that movement were continuous for half and only half of that interval. For the length of the night as determined by the second method, this is on the average not a rash assumption. For the initial and especially for the final active periods, which are included under the first method, the assumption is very great, however, since activity is not limited to half of each interval but is generally scattered over the whole interval. From this point of view, the second method probably gives a more adequate measure of the child's behavior. From the preceding discussion it appears to give also a truer measure of the quietness of the child's sleep. (For some purposes the two measures can be treated alike. The rank-order correlation between MRP1 and MRP2 for 13 children, when each mean is based on from 101 to 291 daily records, is .84; and for 22 children, including the above 13, even when some of the means are based on as few as 21 daily records, it is .95.) The only difficulty here is that we are not certain that the first rest period greater than five minutes coincides consistently with what we ordinarily call sleep. This difficulty could be partly cleared up by an observational study; but since it is due partly to lack of definiteness in our ordinary concept of sleep, it does not constitute a legitimate criticism of the present method.

Summary of Definitions

The length of night is the total time the child is in bed (T_1). When measured by the number of five-minute intervals marked by the kinetograph, a deduction of one interval must be made to arrive at the true length. This has been done in all cases where values of T_1 are given in hours and minutes, but not in case of the T_2 used in the formula for the mean rest period. The error is only about .75 per cent. The length of the night's rest or the length of sleep (T_2) is the period that starts with the beginning of the first rest period greater than five minutes and ends with the child's rising in the morning, or with the beginning of the final active period if the latter is greater than 25 minutes. When it is measured in intervals of five minutes, this value must be corrected by subtracting half an interval. This has been done in all cases where values of T_2 are given in hours and minutes, but not in case of the T_2 used in the formula for MRP. The error is usually less than one half of one per cent.

The initial active period is the initial active five-minute interval or group of intervals. It intervenes between the time of going to bed and the first rest period greater than five minutes. It always includes the first interval on the record, and may include only this one. A correlation of one interval has been subtracted from the means in every table that gives this value in minutes.

If the child spends the last 25 minutes or more of his night in continuous activity (without becoming quiet for more than 2.5 minutes), this continuous activity is called the final active period. If this final period is less than 25 minutes, it is disregarded.

Where T_1 is the number of five-minute intervals in the night as defined above, t_1 the number of those intervals during which movement occurs, and MRP_1 the mean rest period in minutes for all the time spent in bed,

$$MRP_1 = \frac{5\,T_1}{t_1} - 2.5\,.$$

If T_2 as defined above designates the number of five-minute intervals spent in comparative rest or sleep, t_2 the number of these intervals during which movement occurs, and MRP_2 the mean rest period in minutes for the time devoted to sleep,

$$MRP_2 = \frac{5T_2}{t_2} - 2.5\,.$$

A movement, to be taken account of in the present study, must involve a change in bodily position or a shift of body mass equivalent to the movement of an arm or leg or the very sudden movement of a forearm, sufficient to cause a one-millimeter displacement of the record line on Johnson's kinetograph. As the term "sleep" is used in this study, the criterion of sleep in a child aged from two to five years is the cessation of movement for at least five continuous minutes.*

FINDINGS

Findings on distribution of activity are presented in Chapter II. Comprehensive data on the mean rest periods, length of night, length of sleep, and length of initial active period are presented in the table in the Appendix and are summarized in Table 1. The method used in the Appendix of recording the data by months makes it possible for anyone to check or rework them for age, sex, or seasonal factors in any way he may see fit. The ages of the children are also given in the Appendix. Until they were ready for final presentation, the data were kept in the form of sums of measures (N), and sums of measures squared for each month for each child, on tabular work sheets described by the author in a published article (10). Thus these data could readily be recombined into age, season, or sex groups and the means and sigmas easily calculated from the work sheets. Other workers with our data can do the same thing at the cost of some additional labor by using the monthly means and sigmas given in the Appendix and the formula given in reference 10.

SUMMARY

1. This study records the activity during sleep of twenty-two children who were attending nursery school at the Institute of Child Welfare, University of Minnesota, in 1928 and 1929. The purpose of the study was to establish norms of the distribution of night activity and of the mean length of time between successive major changes of position. Changes in the children's bodily position while in bed were recorded by Johnson kinetographs.

2. The nights were divided into periods of five minutes each, and mean rest periods were calculated by two formulas. MRP1 is the average period of inactivity from the time the child went to bed

* For a definition of sleep see Johnson's essay (21; also 27, page 501).

TABLE I.—SUMMARY OF DATA ON INDIVIDUAL SUBJECTS

Subject	No. of Nights	Time in Bed Mean	σ	MRP1 Mean	σ	Time Asleep Mean	σ	MRP2 Mean	σ	Initial Active Period No. of Nights	Mean	σ
1	250	11h 13m	33.5	5.90m	0.74	10h 23m	44.5	6.40m	0.85	252	41.1m	21.3
2	101	11h 33m	42.0	7.90m	1.28	10h 32m	69.0	9.11m	1.52	103	49.9m	31.0
3	280	10h 48m	50.0	7.87m	1.46	10h 10m	51.0	8.65m	1.63	289	23.6m	14.7
4	266	10h 52m	72.0	7.99m	0.58	10h 19m	62.0	8.48m	1.46	268	28.5m	26.5
5	220	10h 00m	55.5	7.58m	1.16	9h 30m	55.0	8.18m	1.19	233	20.9m	20.8
6	267	11h 58m	73.0	7.32m	0.88	10h 58m	62.5	8.31m	1.11	280	43.1m	36.8
7	259	10h 57m	60.5	7.73m	1.24	10h 23m	61.0	8.41m	1.18	294	28.8m	25.9
8	228	11h 15m	48.5	6.44m	0.83	10h 35m	47.0	6.96m	0.96	230	34.9m	16.1
9	200	12h 16m	73.5	7.44m	1.17	10h 52m	65.5	8.82m	1.36	213	48.9m	37.9
10	291	11h 52m	62.0	6.71m	0.87	10h 59m	57.0	7.38m	1.10	298	39.1m	32.2
11	116	10h 33m	76.0	7.91m	1.36	9h 39m	69.5	8.95m	1.40	137	28.3m	26.3
12	284	10h 51m	40.5	7.52m	1.10	10h 10m	42.0	8.23m	1.11	289	28.2m	24.0
22	116	10h 41m	50.0	8.41m	1.24	10h 00m	48.5	9.35m	1.30	120	34.0m	28.5
13	21	11h 32m	11.0	7.68m	1.09	10h 55m	53.0	8.32m	1.19	25	33.4m	23.2
14	57	10h 36m	69.5	7.36m	1.11	10h 05m	57.5	7.79m	1.03	59	24.6m	48.2
15	70	10h 50m	80.5	6.32m	0.82	9h 57m	53.5	7.05m	1.14	72	34.5m	16.5
16	35	10h 45m	29.5	6.39m	0.66	9h 45m	33.0	7.20m	0.82	35	49.5m	23.0
18	67	10h 15m	53.5	6.76m	0.98	9h 35m	49.0	7.29m	1.00	72	22.3m	21.3
19	29	10h 19m	50.5	6.91m	0.81	9h 33m	59.0	7.65m	0.90	32	34.4m	12.4
20	42	10h 06m	58.5	7.33m	0.92	9h 38m	54.0	7.84m	0.89	43	25.1m	15.8
21	66	11h 11m	40.5	7.06m	0.88	10h 27m	45.0	7.75m	1.00	68	35.5m	22.6
32	74	10h 50m	48.0	10.13m	2.15	10h 16m	52.5	11.33m	2.51	74	31.2m	18.3
All	3,339	11h 06m	70.0	7.36m	1.32	10h 20m	63.0	8.12m	1.54	3,486	33.4m	27.9

until the time he got up in the morning. MRP2 is the average period of inactivity during that part of the night between the initial active period before going to sleep and the final active period after waking but before getting up in the morning.

3. Usable records were obtained for a total of 3,339 nights, during which the children were in bed, on an average, 11 hours and 6 minutes, and asleep 10 hours and 20 minutes. MRP1 was 7.36 minutes, MRP2 was 8.12 minutes.

III. DISTRIBUTION OF ACTIVITY
DURING THE NIGHT

Following the method described in Chapter I, we constructed curves for each of thirteen subjects, each curve based on records of one hundred nights.* The curves for three subjects are reproduced here to illustrate the form of the temporal distribution of nocturnal motility and to show individual differences and similarities. When the tabulation of each night's record is started at the time of going to sleep, i. e., with the first rest period greater than five minutes, the curves take the form shown in Figure 4. When the records are tabulated beginning with the time of going to bed, the curves take the form of the samples in Figure 3. The same one hundred nights are used in plotting the two curves for any one subject.

Individual Differences

Inspection of the individual curves shows that they are less different from each other than one would expect from Johnson's curves for adults. The differences are, as a matter of fact, less marked than in curves that were plotted before the data were all tabulated (9, page 176). This brings into question the significance of these individual differences. The question might be formulated as follows: Are the differences between the forms of the curves for different individuals greater or less than the similarities between them? In order to find the answer to this question, it was decided to intercorrelate the motility records on the various subjects. The records were tabulated after the first rest period greater than five minutes, since this period was thought to be more characteristic of sleep than the preceding intervals. If these intercorrelations should approach 1.00 when corrected for attenuation or for the variability of the individual records, this would show that the curves for the different subjects were of the same form, within the limits imposed by the unreliability of the individual measures. If, however, these correlations were very low or approached zero, we should conclude that each individual's curve has a form peculiar to that individual,

* Subjects 1 to 12 inclusive and Subject 22. On other subjects records were incomplete.

and that different subjects do not tend to be motile and quiet at the same relative stages of their rest.

Reliability of individual curves.—In order to obtain reliability coefficients to be used in the correction for attenuation, the data on each subject were tabulated for fifty odd-numbered and fifty even-numbered nights. The number of nights on which activity occurred within any one five-minute interval was expressed as a percentage of the fifty nights, in order that proratings could readily be made for those last few intervals of the composite night which represent records of fewer than fifty nights, and incidentally for the purpose of plotting curves. The number of proratings is small. Such measures of each of the intervals in the composite of odd-numbered nights form a series that is to be correlated with the corresponding series for the even-numbered nights. The number of intervals in the night was limited to one hundred in order to reduce the number of proratings necessary. The same limit was set for all children in order that the correlations might be comparable and that the Hollerith tabulating machine might be used in making the intercorrelations. The odd-even reliabilities were calculated by means of the Anderson correlation chart, class intervals of five units each being used throughout. The coefficients with their probable errors, together with the Spearman-Brown predicted values demanded by the attenuation formula for comparison with the total correlations, are given in Table 2.

TABLE 2.—CORRELATIONS BETWEEN ACTIVITY SCORES ON FIFTY ODD- AND
FIFTY EVEN-NUMBERED NIGHTS FOR INDIVIDUAL SUBJECTS

Subject	r	PE_r	Spearman-Brown ρ	Subject	r	PE_r	Spearman-Brown ρ
1	.608	.04	.756	8	.738	.03	.849
2	.635	.04	.777	9	.656	.04	.792
3	.458*	.05	.628†	10	.653	.04	.790
4	.601	.04	.751	11	.556	.05	.715
5	.583	.04	.737	12	.716	.03	.834
6	.733	.03	.846	22	.713	.03	.832
7	.631	.04	.774	Mean . .	.637		.777

* Corrected from .450 by Sheppard's correction for effect of broad categories.
† This figure would have been .621 if the correction in r had not been made.

Correlations between individuals.—For the intercorrelations, the odd and even series were combined into a total or composite series for each subject. These values were coded into class intervals of five

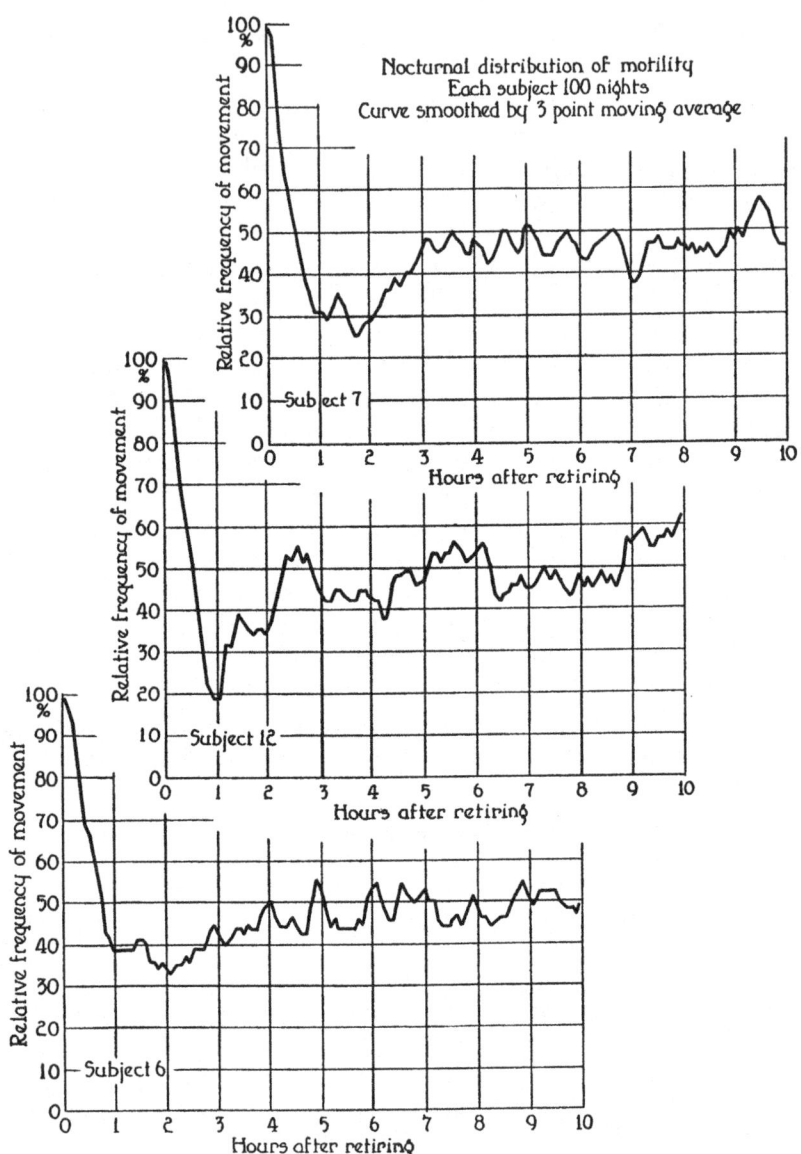

FIGURE 3.—RECORD OF NOCTURNAL MOTILITY FOR THREE SUBJECTS,
TABULATED FROM TIME OF GOING TO BED

18

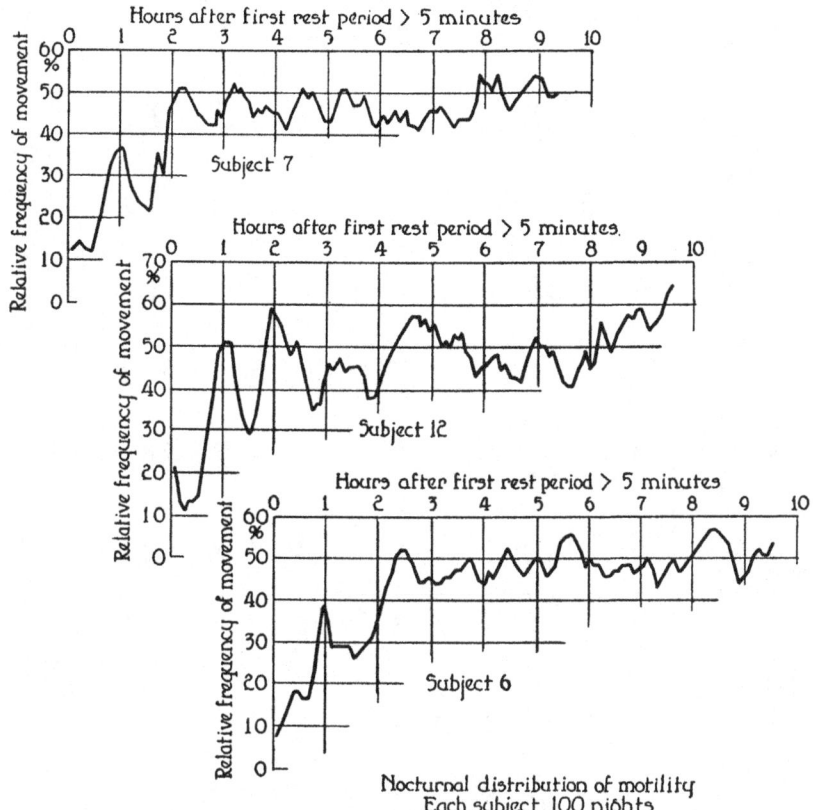

FIGURE 4.—RECORD OF NOCTURNAL MOTILITY FOR SAME THREE SUBJECTS, TABULATED FROM TIME OF GOING TO SLEEP

units each, numbered from 1 to 15. Tabulations were made with the aid of the Hollerith machine, and the coefficients calculated according to the method suggested by Warren and Mendenhall (61). The coefficients, as well as the corrected values, are presented in Table 3. The corrected values were gained by use of Spearman's formula:

$$\text{true } r_{12} = \frac{r_{12}}{\sqrt{r_{1I}} \ \sqrt{r_{2II}}}$$

where r_{12} is the obtained correlation between Subjects 1 and 2, r_{1I} the Spearman-Brown reliability of Subject 1, and r_{2II} that of Subject 2. The means of corrected correlations of each subject with all the others are given merely for purposes of comparison.

TABLE 3.—INTERCORRELATIONS BETWEEN SUBJECTS IN DISTRIBUTION OF ACTIVITY ACCORDING TO TIME OF NIGHT*

(N=13 Children, 100 Nights Each)

Subject	1	2	3	4	5	6	7	8	9	10	11	12	22
1	.756	.754	.674 .684	.669	.677	.704	.719	.724	.537 .542	.761	.623	.664	.774
2	.984	.777	.574 .583	.633	.593	.684	.756	.582	.565 .571	.618	.674	.591	.760
3	.993	.835	.628	.632 .642	.549 .557	.650 .659	.678 .689	.623 .632	.552 .567	.596 .605	.617 .626	.582 .591	.649 .659
4	.888	.828	.934	.751	.646	.757	.792	.759	.705 .713	.662	.618	.650	.698
5	.907	.783	.819	.868	.737	.694	.733	.644	.620 .626	.652	.618	.683	.684
6	.880	.844	.904	.950	.879	.846	.813	.677	.809 .818	.721	.735	.691	.821
7	.940	.974	.988	1.039	.971	1.004	.774	.770	.770 .778	.752	.676	.796	.806
8	.903	.717	.866	.950	.815	.799	.950	.849	.648 .654	.670	.519	.727	.677
9	.701	.728	.804	.924	.820	.999	.994	.799	.792	.690 .697	.626 .632	.590 .596	.727 .734
10	.984	.789	.877	.860	.854	.882	.961	.818	.882	.790	.594	.654	.749
11	.848	.904	.934	.843	.851	.946	.908	.666	.840	.790	.715	.562	.716
12	.836	.734	.817	.822	.871	.823	.991	.863	.734	.805	.727	.834	.718
22	.976	.946	.912	.884	.874	.979	1.004	.806	.905	.924	.928	.862	.832
Mean	.903	.839	.890	.899	.859	.907	.977	.829	.844	.869	.849	.824	.917

* The coefficients above the diagonal of italic figures are the raw correlations, uncorrected for attenuation. The figures below the diagonal are the values corrected for attenuation according to the individual odd-even reliability coefficients in the italic. Where two values appear for any correlation (for instance, in the case of variables 3 and 9, where there are only nine class intervals in the range), the first is the obtained value, while the one below it is the value corrected for the effect of a small number of broad class intervals by the Sheppard formula $\sigma = s - .0833$, where $s =$ obtained σ squared. These corrections, where made, range from .0057 to .0145.

"Time of night" refers not to clock hours but to corresponding five-minute intervals in the nightly rest of the several children.

20

The raw correlations (mean = .65) are of the same order of magnitude as the raw reliability coefficients (mean = .64), and of nearly the same order of magnitude as the Spearman-Brown predicted values (mean = .776). Correction for attenuation raised the intercorrelations so high (range .67 to 1.04, mean .88) that it is evident that the individual factors introduced by the subjects as individuals are less potent in determining the relationships than are the general factors that apply to all subjects. As applied to the distribution of nocturnal motility, this means that the curves are very similar from child to child, rather than peculiar to the individual. The fact that the differential factors are not unimportant, however, especially in the case of some of the children, is shown by the fact that some of the corrected correlations are far below 1.00. The reader should also be cautioned not to apply this conclusion to adult subjects. H. M. Johnson's work clearly shows the importance of individual differences in the curves of adults. When his curves on adults are compared by inspection, the differences are striking. When analogous curves on children are compared, however, the *resemblances* are striking.

The suggestion that this agreement in relative time of motility is due to external conditions is nullified by the fact that the children were sleeping in widely separated homes and by the fact that the curve of motility is not based on clock hours, but begins with the beginning of each child's own nightly rest, which varies from child to child and from night to night.

The Group Curve

The conclusion stated above led us to make a composite curve of motility, combining the data on all thirteen children. If the conclusion were not true, this curve would tend to be a straight line, the individual differences canceling each other and producing a composite or average curve with no definite fluctuations. The fluctuations in Figure 5 are obvious.

There is an apparent decrease in the definiteness and magnitude of the fluctuations as the night progresses. This is just what we should expect when we remember that this curve is based upon records of 1,300 nights of varying lengths. By definition, the starting point or first five-minute interval of each night is comparable with that of every other night. Because the nights are of different lengths, however, the corresponding points on different nights become less and less comparable as we proceed toward morning. For example, all the ninetieth intervals from the beginning are thrown together, to be sure, but the tenth from the last interval on the composite curve is by no means composed of the tenth from the last interval on each of the nights. Similarly, the last interval on the composite curve represents few component nights, since nights differ in length. The same argument applies in a lesser degree to the curves for single subjects. The remedy for this situation would have been to fractionate every night into the same number of equal parts, regardless of

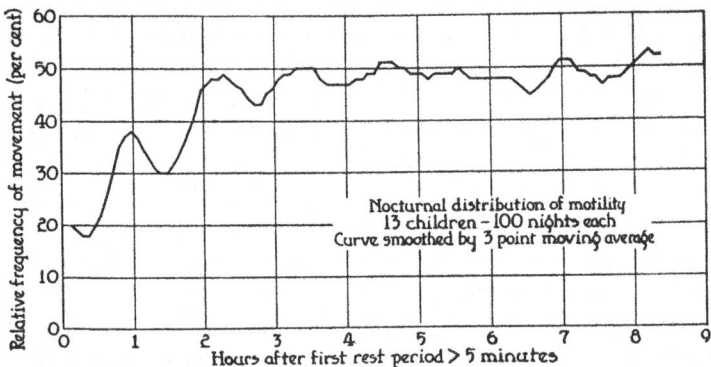

FIGURE 5.—COMPOSITE RECORD OF NOCTURNAL MOTILITY FOR THIRTEEN SUBJECTS

their length, instead of into equal (five-minute) parts of varying number. Each fractional part of any night would then be comparable with the corresponding part of any and all other nights. Composite curves would have the same meaning throughout their length with respect to the night as a whole, rather than with respect to its beginning only. This method is similar to the treatment of learning curves suggested by Vincent (60, pages 16–17), described also by Hunter (20, pages 567–68) and Kjerstad (33, page 26), and used by Robinson and Heron (46, pages 428–48) and others.

The effect of home conditions.—The question arises whether individual variations from the group curve are due to individual characteristics of the various children or to home conditions. Do the conditions under which a child sleeps force his curve out of the form that is taken by the composite curve for the other children, or is the form of the curve independent of the home environment?

This question was attacked in the following manner. On the basis of brief descriptions of the homes written by the experimenter, who visited the homes almost daily for several months and who was also present as a guest on a number of evenings, ten judges ranked the home conditions in the order in which they might be least expected to disturb the normal, spontaneous course of sleep activity. These judges were graduate students and members of the staff of the Department of Psychology and the Institute of Child Welfare. Their rankings were averaged, the averages ranked in order, and the new ranks correlated with the rank order of the average intercorrelations on the form of the curve. (See Table 3.) This gives us a rank-order correlation between the home conditions as ranked by the judges and the extent to which a child's motility curve agrees with that of his fellows. The correlation of .54 ± .14 means that the children who had the poorest home conditions tended to differ most from the rest of the group in distribution of activity during sleep. In other words, the poorer a child's sleeping conditions were, the less closely his curve fitted the group curve.

The differences between individuals, then, are affected by conditions external to the organism, and are not entirely "individual differences." The correlation is so far from perfect, however, that the conditions with which we are dealing cannot be regarded as causing the entire difference. True individual differences probably do affect the form of the activity curves. Nevertheless, it should be remembered that these differences are exceedingly small in comparison with the *similarity* between the individuals in the group, and very much smaller than the individual differences found among adults.

Reliability of judgments.—The rank-order interrelations among the ratings of the ten judges range from .71 to .98, with a mean of .87. This shows that the homes were rated reliably, as is shown also by the fact that the averages range from 1.5 to 13, thus preserving almost the entire original range (1 to 13), with very little regression toward the mean; and by the close agreement between the average ranking for each child and the rank order of these averages. Thus the average of the individual rankings and the rank order of this average, respectively, for Subject 1 are 1.6 and 2.0; for Subject 2, 8.0 and 9.0; and for the other children in numerical order, 6.0 and 5.0, 6.9 and 8.0, 4.4 and 3.0, 10.5 and 11.0, 1.5 and 1.0, 6.1 and 6.5, 13.0 and 13.0, 6.1 and 6.5, 11.9 and 12.0, 10.4

and 10.0, 4.6 and 4.0. (The last is for Subject 22.) The average rank of 13.0 for the home of Subject 9 shows perfect agreement of the ten judges upon this home; the average rank of 11.9 for that of Subject 11, only one dissenting judge. The fact that the correlation between the average ranking of the homes and the agreement of the individual subjects with the group motility curve is so much smaller than the reliability of the judgments adds weight to the statement that home conditions do not entirely explain the individual differences. The descriptions on which the judges rated the homes are given in the Appendix.

RELIABILITY OF DIFFERENCES BETWEEN CORRELATION COEFFICIENTS CORRECTED FOR ATTENUATION

On the basis of the average correlation between each individual and all the others, the group was divided into two sections, one containing the children with the higher average correlations and the other those with the lower ones. The correlations in each division are presented in Tables 4 and 5. The largest difference between the highest and the lowest correlation for either group or both groups is .37; the largest difference between subjects in different groups is .30; and the largest difference between the relationships in the group as a whole is .37.

The sigma, $\sigma r_a - r_b$, of the difference between two correlations corrected for attenuation is calculated from the formula

$$\sigma d = \sqrt{\sigma^2_1 + \sigma^2_2},$$

which in this case becomes

$$\sigma r_a - r_b = \sqrt{\sigma r^2_a + \sigma r^2_b}.$$

The sigmas of the corrected correlations σr_a, σr_b, are calculated from Kelley's revision

$$\sigma r_{\infty\infty} = \frac{r_{\infty\infty}}{\sqrt{N}}\left(r_{\infty\infty}^2 + \frac{1}{r^2_{12}} + A_{11} + A_{211}\right)^{1/2}$$

of Spearman's formula, where $r_{\infty\infty}$ is r_a or r_b, and where A and $r^{1/4}$ are taken from Kelley's table (32, pages 208–11).

None of the differences within either division are statistically significant, but some of the differences not so confined are significant (at least three times their own sigma). Composite curves are presented in Figure 6 for the two divisions separately, and the composite curve for the total group in Figure 8, page 29.

TABLE 4.—CORRECTED INTERCORRELATIONS IN DISTRIBUTION OF ACTIVITY BETWEEN SUBJECTS HAVING HIGH AVERAGE CORRECTED INTERCORRELATIONS WITH THE WHOLE GROUP

Subject	Subject						
	1	3	4	6	7	10	22
1		.99	.89	.88	.94	.98	.98
3	.99		.93	.90	.99	.88	.91
4	.89	.93		.95	1.04	.86	.88
6	.88	.90	.95		1.00	.88	.98
7	.94	.99	1.04	1.00		.96	1.00
10	.98	.88	.86	.88	.96		.92
22	.98	.91	.88	.98	1.00	.92	
Mean.	.94	.93	.93	.93	.99	.91	.95

TABLE 5.—CORRECTED INTERCORRELATIONS IN DISTRIBUTION OF ACTIVITY BETWEEN SUBJECTS HAVING LOW AVERAGE CORRECTED INTERCORRELATIONS WITH THE WHOLE GROUP

Subject	Subject					
	2	5	8	9	11	12
2		.78	.72	.73	.90	.73
5	.78		.82	.82	.85	.87
8	.72	.82		.80	.67	.86
9	.73	.82	.80		.84	.73
11	.90	.85	.67	.84		.73
12	.73	.87	.86	.73	.73	
Mean.	.77	.83	.77	.78	.80	.79

RHYTHMIC FLUCTUATIONS IN THE ACTIVITY CURVES

The composite curve for the higher division shows greater fluctuations than that for the total group. It also shows fluctuations as great as or greater than those in the curve for the lower division, in spite of the tendency for a larger number of subjects to give smaller fluctuations on account of the averaging effect. These fluctuations in the curve for the higher division are not only greater in magnitude but are also more clearly defined than those of the lower division, as we should expect from the fact that division was made on the basis of the intercorrelations. Disregarding this difference in definiteness and picking out the crests and troughs of the waves as best we can, we may count the number of half-waves and measure their lengths and variabilities. For the higher division we find thirteen half-waves having a mean of 7.3 five-minute units and a mean deviation of 1.9. (See Figure 6.) For the lower division we find thirteen half-waves with mean 7.3 and mean deviation 1.7. On the total group curve we find thirteen such waves with mean 7.3 and mean deviation 1.7.

The consistency of these fluctuations from one curve to another, both in number (and hence in mean length) and in variability of length, shows that they tend to be periodic for all the children taken in groups. The periodicity or rhythmic character of these curves can best be observed directly by the reader. The mean half-wave of 7.3 means that the child tended to become more and more restless for about thirty-five minutes and then to become more quiet for about the same length of time. This fluctuation may be thought of as due to a gradual accumulation of pressure, temperature, and tactile stimuli demanding a change of posture for their relief, after which the subject becomes gradually more quiet.

The question arose whether this process of becoming quiet was more rapid than the phase of the wave indicating increasing restlessness. That it was not, at least for these subjects, is shown by comparison of the mean crest-trough and trough-crest half-wave lengths, which are respectively 7.3 and 7.3 for the higher division, 7.7 and 6.9 for the lower division, and 6.9 and 7.7 for the total group. Thus the only apparent tendency is for the phases to be the same. The fluctuation in frequency of activity might, on the other hand, be thought of in terms of Pavlov's concept of sleep as inhibition, the periodicity indicating a more or less constant limit of endurance or maintenance of the inhibition of movement. Whatever the implications of the fact may be, there seems to be some constancy in the lengths of the fluctuations, so that the curves of nocturnal distribution of motility become rhythmic. This is substantiated by the findings of Michelson (40), of de Sanctis and Neyroz (49), and of Johnson and Weigand (26, page 15). Howell (19) found similar fluctuations in the volume of the arm during sleep. Mönninghoff and Piesbergen (42) and Czerny (4) found fluctuations but no rhythms. These studies are compared by Johnson and Swan (26, pages 15ff). Our curve is very different from that of Kohlschütter (35), which, however, is erroneous (cf. 54; also 26, pages 13–15).

SEX DIFFERENCES

When the group is divided on the basis of sex, the curve for girls shows half-waves with a mean of 7.3 and a mean deviation of 1.6. The curve for boys shows a mean of 7.3 and a mean deviation of 2.3. The dissimilarity between these sex curves is even greater than that between curves of the two divisions made on the basis of the intercorrelations, as can be seen by comparison of Figures 6

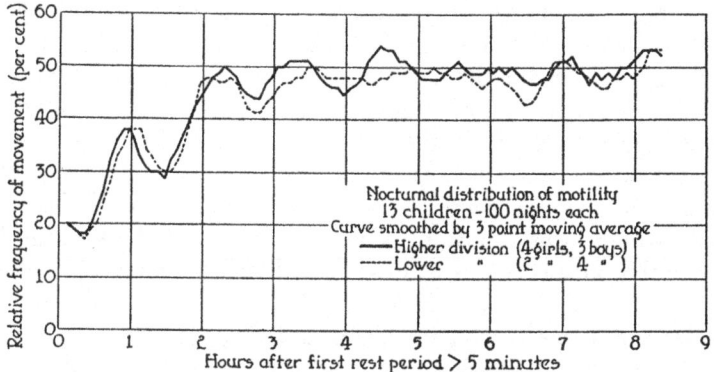

FIGURE 6.—COMPOSITE RECORD OF NOCTURNAL MOTILITY FOR SEVEN CHILDREN SHOWING HIGHER AVERAGE CORRELATIONS AND SIX CHILDREN SHOWING LOWER AVERAGE CORRELATIONS WITH MAIN EXPERIMENTAL GROUP

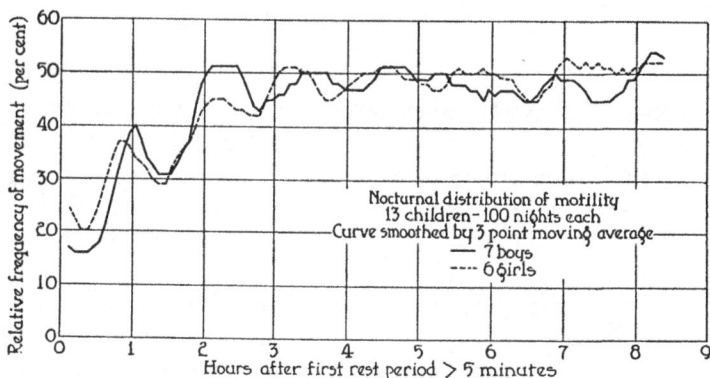

FIGURE 7.—COMPOSITE RECORD OF NOCTURNAL MOTILITY FOR THE SEVEN BOYS AND SIX GIRLS IN MAIN EXPERIMENTAL GROUP

and 7. This points toward a sex difference, which is small, however, and may not be significant. The mean heights of the sex curves differ by less than half a point, or half of 1 per cent of the number of measures on either group considered in each five-minute interval. The proper test for a sex difference in the form of the motility distribution curves, however, is the comparison of like-sex and unlike-sex correlations. Such comparison shows that the average of the like-sex correlations is .90 as against the average of .86 for the unlike-sex correlations and .88 for the average of the total group. Furthermore, the average of the correlations of boys with boys and the average of girls with girls are both higher than the average for the total group. The average for the boys is .90 and for the girls .91 as against .88 for the total group. Still further, the average of each child with all the other children of his own sex is greater than his average with all the children of the opposite sex in every case except two, in which there is no difference. There are no reversals. The averages of the correlations of each child with all the others of his own and the opposite sex are: for Subject 1 the average like-sex correlation was .96 as against an unlike-sex average of .87, and for the other subjects in numerical order the like-sex and unlike-sex correlations, respectively, were .89 and .80, .91 and .88, .93 and .87, .87 and .85, .91 and .91, .99 and .95, .86 and .80, .88 and .81, .87 and .87, .88 and .83, .85 and .80, and .94 and .90. The significance of these differences has not been computed statistically, but the trend is so consistent as to be fairly conclusive evidence of a small sex difference in the form of the activity curve.

Form of the Activity Curve for the Entire Night

In order to show the distribution of activity beginning at the time of going to bed, a composite curve for thirteen children has been plotted. The curve is a composite of one hundred nights' records on each child, each record starting at the time the child went to bed. Thus the initial active period of each night, which was excluded by the other method, is now included. This curve (Figure 8) is made seven units longer than the other (Figure 5) in order to cover all the same activity and also the initial active period, which averages about seven intervals or thirty-five minutes; see Table 6. (The actual average would be nearer thirty-two minutes when it is calculated to the fraction of an interval and correction made for the statistical method, as is done for all the records on all twenty-two children in Table 1. For the present purpose, however, the present method is sufficient.) The first point on this curve is 100 per cent by definition, just as the first point (which, however, was not plotted) on the other curve was zero. The rapid drop in the curve is due to the fact that the initial active period of the various records was of varying length, and that the relative frequency of movement was low immediately after this period, as shown by Figure 5.

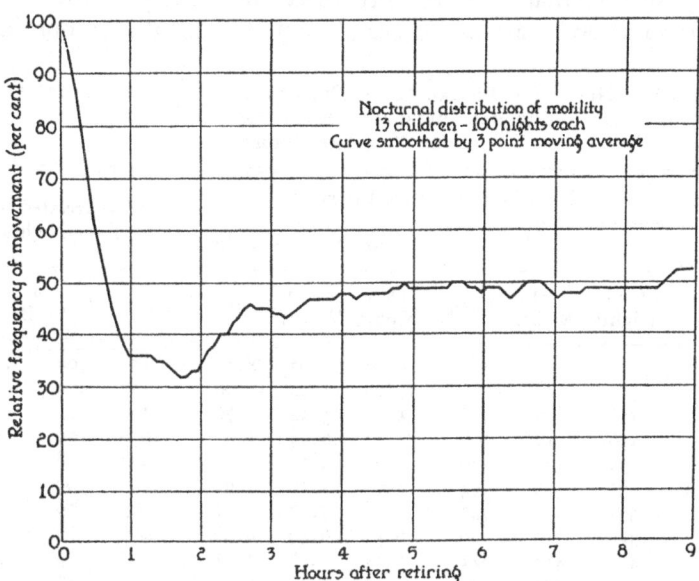

FIGURE 8.—COMPOSITE RECORD OF NOCTURNAL MOTILITY FOR THE THIRTEEN CHILDREN IN THE MAIN EXPERIMENTAL GROUP, TABULATED FROM TIME OF GOING TO BED

This suggests that the initial active period represents activity of a different kind from that represented by the rest of the curve. Inspection of the curve (Figure 8) shows that the measures on the first seven intervals (the average initial active period) are of a different order of magnitude from the rest. It has already been suggested (page 10) that the initial active period, like the final active period, is a record of waking activity, whereas the rest of the curve is a record of sleep activity. The rapid drop in the relative frequency of movement from 100 per cent to a much lower level, and the general maintenance of this level, support such a view.

This view is also supported by the fact that Kreidl and Herz

(36) observed that the time of going to sleep coincided with the end of an initial active period, similar in length to the one with which we are now dealing. It also finds some support in the study of Foster, Goodenough, and Anderson (7), which indicates that children take about twenty minutes to go to sleep. It finds more support, however, in the conclusions of Boynton and Goodenough (3), who worked with the same children used in the present study. The mean length of time that they find is required for nursery school children to go to sleep is almost identical with the length of our initial

TABLE 6.—NUMBER OF INITIAL ACTIVE PERIODS (FIVE-MINUTE INTERVALS) OF
INDIVIDUAL SUBJECTS ON FIFTY ODD- AND FIFTY
EVEN-NUMBERED NIGHTS

| Subject | No. of Initial Active Intervals | | | | | | Odd Correlated with Even Nights | | |
| | Total | | | Mean No. per Night | | | | | |
	Odd Nights	Even Nights	Both	Odd Nights	Even Nights	Both	r	PE_r	Spearman-Brown ρ
1 . . .	475	401	876	9.50	8.02	8.76	.156	.09	.270
2 . . .	554	552	1106	11.08	11.04	11.06	.127	.09	.225
3 . . .	299	277	576	5.98	5.54	5.76	−.140	.09	−.246
4 . . .	256	312	568	5.12	6.24	5.68	.059	.09	.111
5 . . .	190	176	366	3.80	3.52	3.66	.422	.08	.594
6 . . .	431	563	994	8.62	11.26	9.94	.243	.09	.391
7 . . .	391	351	742	7.82	7.02	7.42	.063	.09	.119
8 . . .	326	320	646	6.52	6.40	6.46	−.076	.09	−.141
9 . . .	407	414	821	8.14	8.28	8.21	.185	.09	.312
10 . . .	445	478	923	8.90	9.56	9.23	.000	.10	.000
11 . . .	390	257	647	7.80	5.14	6.47	−.008	.10	−.016
12 . . .	363	302	665	7.26	6.04	6.65	.002	.10	.004
22 . . .	416	366	782	8.32	7.32	7.82	.040	.10	.074
Mean.	382	367	747	7.65	7.34	7.47			

active period. The difference between the twenty minutes found by Foster and the thirty-five minutes found by Boynton may be explained in part by the larger number and greater heterogeneity of cases used by Foster, and by the fact that Boynton was working with diurnal naps instead of night sleep, but it is probably related also to the fact that Boynton's observations were all made and recorded by a competent and careful observer, whereas Foster's records were made by the parents of the children. It is very probable that there is a tendency for parents to assume that a child is asleep as soon as they can leave him. This would tend to produce the result described.

If our supposition is correct, the proper way to plot the curve of sleeping activity is to take the records from the end of the initial active period, i. e., from the beginning of the first rest period greater than five minutes, as was done in Figure 5. That it is correct is attested by the fact that inclusion of the initial active period in Figure 8 destroys the rhythmic fluctuations of Figure 5. This means that starting all the records with the first rest period greater than five minutes makes corresponding points on the records comparable with reference to the child's night's sleep considered as a biological entity, while starting all the records at the time the child goes to bed does not make them so comparable.

Unreliability

The effect of considering the total time in bed, regardless of sleep, upon the odd-even reliability coefficients for the individual children has some bearing here. Reliability is computed in much the same way as under the other method. From the preceding discussion it is obvious that the inclusion of the first seven intervals would introduce a spurious element since they do not belong in the same logical or biological category as the others, and since in many cases they are of such order of magnitude as not to form a continuous frequency distribution with the other measures considered as variates. The detachment of the initial seven intervals from the bulk of the measures is striking, though it is not found in case of all subjects and particularly not in the case of Subject 5. (See Table 6.) This discrepancy is due in part to the fact that the mean initial active period for this subject is less than four intervals in length instead of seven—the shortest in the group. We should expect, therefore, that the first seven measures would not detach themselves from the rest but would overlap; and in fact we do find that this subject's record shows the greatest overlapping in the group. The converse holds for Subject 2, who has the lowest mean initial active period and the greatest detachment of the first seven measures (no overlapping at all on either variable). This finding provides further evidence that the initial active period is a period of waking activity.

The reliability coefficients, their probable errors, and the Spearman-Brown predicted values are shown in Table 7 along with the values that would have been obtained had the initial seven intervals been included. The values in the first three columns may be compared with those of Table 2. The effect of taking records from the time of going to bed was to reduce the mean reliability coefficient from .64, as found by the previous method, to .43, and to increase the variability. In Table 2 the range of correlations is from .45 to .74, a difference of .29, while Table 7 shows a range of from .25 to .67, a difference of .42. Even if the initial seven measures are included, the mean coefficient is .63, less than when the initial active

TABLE 7.—CORRELATIONS BETWEEN ACTIVITY SCORES ON FIFTY ODD- AND FIFTY EVEN-NUMBERED NIGHTS

Subject	Initial Active Period Excluded (100 Intervals)			Initial Active Period Included (107 Intervals)		
	r	PE_r	Spearman-Brown ρ	r	PE_r	Spearman-Brown ρ
1	.603	.04	.752	.749	.03	.856
2	.501	.05	.668	.661	.04	.796
3	.281	.06	.439	.635	.04	.777
4	.320	.06	.485	.608	.04	.756
5	.337	.06	.504	.425	.05	.596
6	.274	.06	.430	.503	.05	.669
7	.247	.06	.396	.568	.04	.724
8	.669	.03	.802	.773	.02	.872
9	.302	.06	.464	.664	.03	.798
10	.327	.06	.493	.643	.04	.783
11	.552	.04	.711	.693	.03	.819
12	.312	.06	.476	.612	.04	.759
22	.509	.05	.675	.692	.03	.818
Mean. . . .	.426		.561	.633		.771

period is excluded from each record, and the range is greater (from .43 to .77, a difference of .34). This means that in order to get the maximum stability and reliability we must measure each night beginning with the first rest period greater than five minutes rather than with the time the child is put to bed. In the calculations of mean rest periods, this would necessitate the use of what we have called the second method (MRP2).

SUMMARY

1. Sample curves are presented of the distribution of activity over the total time spent in bed and over the time asleep. The odd-

even reliability of the individual curves for the time asleep, based on one hundred nights' records, ranges from .45 to .74 (average .64), and when corrected by the Spearman-Brown formula, from .63 to .85 (average .78).

2. The curves on thirteen children, when intercorrelated by use of the Hollerith tabulator, correlate from .52 to .82 (average .65). These values, when corrected for attenuation, range from .67 to 1.04 (average .88).

3. The size of these intercorrelations makes it evident that the factors introduced by the subjects as individuals are less important in determining the trend of the motility curve than are the factors that influence all subjects. This means that the similarities between the subjects in the matter of active and inactive times of night are very great, whereas the individual differences are relatively small. In this respect children differ from adults, among whom individual differences are large.

4. The size of the corrected correlations mentioned in conclusion 2, while sufficient to justify conclusion 3, does not rule out all individual differences. To some extent these differences are due to unfavorable sleeping conditions in some homes, but this factor does not entirely account for even the small individual differences obtained.

5. The fluctuations in the group curves of activity during sleeping time are periodic or rhythmic. This is shown by the small deviations of the wave lengths as compared with the mean length of the waves and by the consistency of the fluctuations from one curve to another when the group is divided on the basis of sex or even on the basis of the size of the intercorrelations. These rhythms confirm the findings of other authors, most of whom did not, however, have sufficient data to establish their rhythms conclusively.

6. When curves are drawn for the sexes separately, the differences between them are even greater than the differences between the curves for the high- and low-correlation divisions of the group. The validity of the divergence between the sex curves is established by the consistency, throughout the group, of the differences between the average like-sex and unlike-sex intercorrelations for each child.

7. A composite curve was drawn for the records of thirteen children on one hundred nights, but each record was started with the time of going to bed instead of with the first rest period greater than five minutes. This curve has only a bare suggestion of the

fluctuations of the other curve, and has no rhythmic character at all. The individual curves plotted in this way have much lower and more variable reliabilities than those beginning with the first rest period greater than five minutes. This finding supports our policy of treating the data as if the beginning of the first rest period longer than five minutes were the beginning of sleep, and using the measure MRP2 rather than MRP1.

IV. THE MEAN LENGTH OF TIME
BETWEEN MOVEMENTS

From Table 1 on page 14 it can be seen that the mean rest period for the group of children studied is about 7.25 minutes when calculated on the basis of the entire time in bed (MRP1) and about 8 minutes when calculated on the basis of sleeping time only (MRP2). These figures are much smaller than those given by Johnson and his coworkers (24, 25) for college men and middle-aged adults, whose mean rest periods range from about 10 to about 15 minutes, depending on individual and occupational differences. Our findings indicate that children are much more active than adults. Children likewise take longer than adults to go to sleep, when the initial active period is used as a measure. Our value for this agrees in general with figures obtained by other workers (3, 7). For length of sleep and length of stay in bed, also, our figures are similar to the values obtained by other students working with children of the same age (2, 7).

GROUPED AND UNGROUPED DATA

The mean length and standard deviations of each child's mean rest period were obtained by two methods, using first grouped data, based on intervals of .50 minutes, and then ungrouped data, using intervals of .01 minutes. The excellent approximations obtained by the grouping method where there are more than one hundred cases in the distribution show that these values can be accepted with great assurance for many practical purposes. The rank-order correlations between the two methods when each value is based on one hundred or more measures (the values for the first thirteen subjects in the table ranked) are 1.00 for the means and .99 for the sigmas. When the values for all twenty-two subjects are ranked, the correlations are 1.00 for the means and .97 for the sigmas.

The means and sigmas as determined by the grouping method were used in plotting the histograms of frequency distributions and the normal frequency curves which are superposed upon them in Figure 9. The values of the ordinates of the normal curves are calculated from the area values in Table XI of Holzinger (18).

35

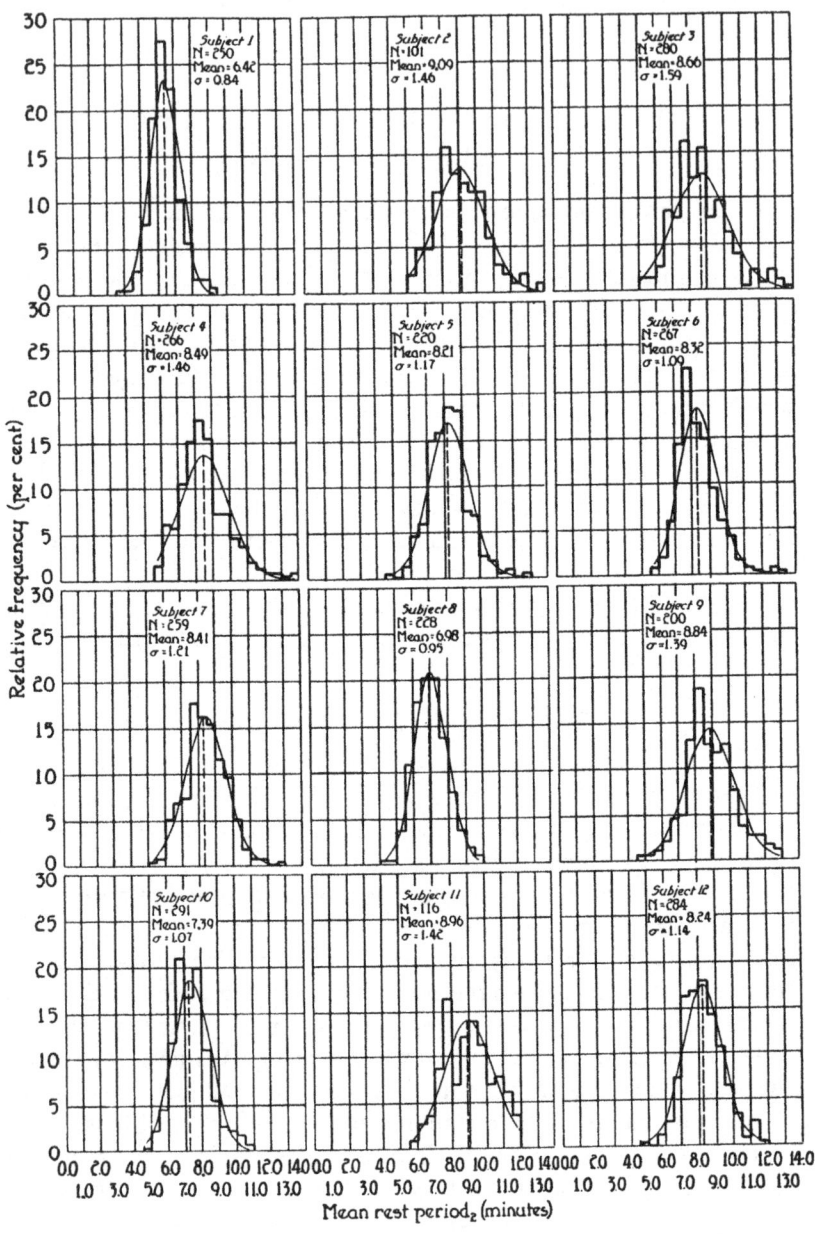

FIGURE 9.—NIGHTLY MEAN REST PERIODS FOR TWELVE SUBJECTS, WITH
NORMAL CURVES SUPERPOSED

Relationship between Length and Quietness of Sleep

The question of the relationship between the length of sleep and its quietness has long been an open one. Czerny's findings (4, page 18) on naps suggest, as Johnson and Swan (26) point out, that the more time the child devotes to rest, the more profound will his rest be. What Czerny found was that during most of the night the child was harder to wake if he had taken a nap that afternoon. Johnson and Weigand (23) found that the longer their adult subjects stayed in bed, the more quietly they rested while they were there.

Individual differences.—In the present study we correlated the length of the rest period for each child, which is a measure of the quietness of sleep, with the mean length of sleep, T_2. Only two of these values are four times their own probable error, and of these one is positive and the other negative. In the case of the other children there is little or no relationship. The findings on only one of our children agree with Weigand's on college students. The fact that we find a positive correlation 4.5 times its probable error on one child and a negative value 7.5 times its probable error on another child indicates that there are individual differences in this relationship at the preschool age, although there may be no such differences at the age level studied by Weigand. We find a rank-order correlation of .12 ± .15 ($N = 21$) between the mean measure of quietness of sleep for each child and the mean length of his night sleep (MRP$_2$ correlated with T_2). This value is too insignificant to indicate anything but a lack of relationship. In other words, as is shown also by the data just presented, there is very little tendency for the more quiet sleepers to be the ones who sleep longer.

The difference between our finding and that of Czerny may well be due to one or more of the following factors: (1) a much larger number of children were used in the present study; (2) many more records were taken on each child in this study; (3) we used night sleep as a measure of length of sleep, whereas Czerny used the afternoon nap as related to a longer total sleep; (4) Czerny measured the magnitude of a discrete stimulus necessary to cause his subjects to stir, while we measured the frequency of "spontaneous" stirrings; and (5) the electrical strength of Czerny's primary current was not necessarily proportional to the stimulus intensity imparted by the secondary current, or even to the strength of the secondary current itself, a circumstance which has been adequately dealt with by Johnson and Swan (26, pages 8–10).

The nap and the night sleep.—The question of naps, especially the effect of an afternoon nap on the quietness of a child's sleep that night, has long been a matter of interest. The findings of Czerny (4, page 18) indicate that an afternoon nap is followed by deeper sleep than that which follows a napless afternoon. It is with some reluctance, therefore, that we suggest the opposite conclusion on what we consider meager evidence. A study like the present one would rightly be expected to deal adequately with this problem. But through a misunderstanding on our part when the study was planned, we expected the daily records of the nursery school to tell us on which days each child took a nap in the nursery school and on which days he did not. We discovered too late that the recording of this item was discontinued some time before our study was started and not resumed till after all our data had been collected.

Czerny's method was questionable, as has been pointed out (26, pages 8–10), and his number of cases was small. We shall therefore present what evidence we have. We have a total of forty-one records on ten children taken on nights known to have followed afternoon naps and a total of eleven records on eight children, taken on nights following days when the nap is known to have been omitted. When the mean rest period (MRP2) of each child for these nights is compared with the norm for that child, we find the following: Of the ten comparisons, six differences point to less quiet sleep after the nap and four to more quiet sleep, but these four differences are all extremely small. One of the first six differences just referred to (for Subject 3) is 1.04 times its own sigma, while the most significant difference in the opposite direction is only .59 times its sigma.

Of the eleven days on which we know definitely that there was no nap, six show a night sleep more quiet than the means (norms) for those respective children, four the reverse, and one practically no difference. The balance of evidence indicates a tendency for sleep on nights following afternoon naps to be more active than the average and on nights not following a nap to be more quiet.

When we turn to the length of sleep and length of stay in bed we find nothing consistent. For the length of stay in bed (T1) there are five differences in one direction and five in the other, none of them significant. For the length of sleep (T2) there are seven differences pointing to longer night sleep after the nap and three to shorter sleep. None of these are significant.

Bedtime.—If the old belief is valid that children should be put to bed early because sleep is more quiet before midnight, we should expect to find that the earlier our subjects were put to bed, the more quietly they rested. Such a finding would produce a negative correlation between the time of going to bed and the mean rest period. From Table 8, which gives the correlation for each of thir-

TABLE 8.—CORRELATIONS BETWEEN TIME OF GOING TO BED AND LENGTH OF MRP2

Subject	No. of Nights	Mean Bedtime (o'clock P.M.)	SD	r	PE$_r$
1	247	7:12	20.5m	−.089	.04
2	101	7:17	29.4m	.126	.07
3	268	7:32	32.1m	.062	.04
4	259	7:58	41.9m	.162	.04
5	213	8:15	26.2m	−.226	.04
6	251	7:21	30.1m	.145	.04
7	248	8:12	20.3m	.216	.04
8	208	7:56	31.0m	−.099	.05
9	175	7:08	41.2m	−.114	.05
10	288	8:13	25.7m	−.136	.04
11	105	8:02	29.0m	.102	.06
12	261	8:27	20.0m	.144	.04
22	111	8:20	24.4m	.064	.06
All		7:50	27.2m		

teen children, we see that only a few of the correlations are large enough to be significant, and they do not agree in direction. The rather even fluctuation of the correlations about zero lends no support to the common opinion. On the contrary, it indicates that the time a child is put to bed has very little to do with the quietness of his sleep. The rank-order correlation of —.03 ± .18 between each child's mean rest period for all nights and his mean time of going to bed supports the same conclusion. It indicates that the differences among the children in usual bedtime probably have nothing to do with their differences in mean rest period. These findings agree with those of Johnson and Weigand (28) on adults, some of whom went to bed at 10:00 and some at 11:30.

The mean bedtime and the variability of the bedtime are also given in Table 8. Since the differences among the children are so great, the mean bedtime given for the group is the unweighted mean, and the sigma for the group is the variability of the individual means from the group mean. The average bedtime for this group is 7:50 P. M. The Institute of Child Welfare of the University of Minnesota found the average bedtime to be 7:48, 7:50, and 7:53 for two- three- and four-year-old children, respectively (2).

MEAN LENGTH OF SUCCESSIVE REST PERIODS

Following the method of Boynton and Goodenough (3) as nearly as the data allowed, we have determined the mean length of the first ten successive rest periods on each of one hundred nights for

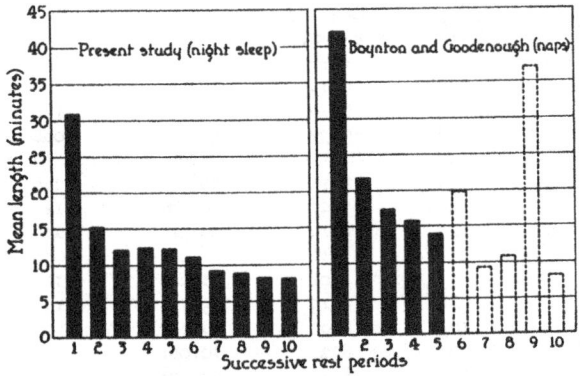

FIGURE 10.—MEAN LENGTH OF SUCCESSIVE REST PERIODS IN PRESENT STUDY AND IN STUDY OF DAYTIME NAPS BY BOYNTON AND GOODENOUGH

each of thirteen children, beginning always with the first rest period longer than five minutes. The mean lengths of successive rest periods for all the children taken together are given in Table 9, along with the corresponding values recorded for naps by Boynton and Goodenough. The same material is shown graphically in Figure 10, in which the outlined bars represent values that the authors consider unreliable because of the small number of cases involved. In so far as the two studies can be compared, the agreement is striking, considering the differences in the conditions under which they were made. The most important differences are that (1) our study utilizes instrumental recording of activity, thus including many shifts in body mass that would not be classed as changes in posture under Boynton's observational method; (2) the unit of time used in the present study is the five-minute interval, whereas the unit used in the other study is one minute; and (3) the present study is based on night sleep whereas the other is based on afternoon naps. All these differences would reasonably be expected to produce the longer rest periods recorded in the Boynton study. One condition was peculiarly controlled in that Boynton took observations on the

TABLE 9.—MEAN LENGTH OF SUCCESSIVE REST PERIODS AS FOUND BY BOYNTON
AND GOODENOUGH (3)* AND BY THE PRESENT STUDY†

	Present Study			Boynton and Goodenough	
Rest Period	Mean No. of Minutes	σ	No. of Naps	Mean No. of Minutes	σ
1	30.9	20.7	278	42.0	24.3
2	15.1	16.4	231	21.7	16.8
3	12.1	14.3	168	17.4	14.7
4	12.5	13.2	100	15.7	12.2
5	12.2	14.2	53	14.1	11.1
6	11.1	13.9	27	19.8	13.4
7	9.2	11.3	7	9.1	5.3
8	8.9	11.8	4	10.8	6.5
9	8.4	10.9	2	37.0	15.0
10	8.2	11.0	1	8.0	

* Boynton and Goodenough's data were based on daytime naps.
† Based on records of one hundred nights for each of thirteen children.

same children (among others) as are included here. The two studies show the same general though not necessarily gradual decrease in the length of rest period as sleep progresses. For the subsequent course of the distribution of activity the reader is referred to Figure 5. Boynton and Goodenough see a relationship between increase in frequency of movement and Miles's (41) reported increase in mean hourly loss of weight through perspiration as sleep continued. We suggest that our curve of distribution of motility (Figure 5) is better suited to this comparison, since it is based on equal time units.

SUMMARY

1. The mean rest period for the whole time in bed averaged between 7 and 7.5 minutes for all the children. There are large individual differences, however, the normal range of individual averages being from 5.9 to 8.4 minutes.

2. The mean rest period during the time devoted to sleep was

about 8 minutes. The averages for normal subjects ranged from 6.4 to 9.4 minutes.

3. The values in conclusions 1 and 2 are to be compared with values for individual adults ranging from 6 to 22 minutes, and with averages of from 7 to 13 minutes, the differences depending apparently upon occupational differences. The children were noticeably more active in their sleep than are adults, moving almost twice as frequently on the average.

4. The length of initial active periods shows that children require almost twice as long to become quiet or to go to sleep as do adults. The average for children is between 30 and 35 minutes and for adults between 10 and 20 minutes.

5. When the length and quietness of sleep were correlated, some significant individual correlations but no group tendencies were found. For most of the subjects very low correlations were obtained, but those that are reliable indicate that there are individual differences, some children sleeping more quietly when they sleep longer and some more restlessly.

6. A small number of data suggest that the afternoon nap may cause the subsequent night sleep to be more active.

7. The hour at which the child goes to bed has no effect upon the quietness of his sleep. Neither do the children who habitually go to bed early sleep more quietly than those who go to bed later. The hours before midnight are no better for sleep than the hours after midnight.

8. The mean length of the first ten successive rest periods in the night sleep was determined after the manner suggested by Boynton and Goodenough, and our findings are similar to theirs. The mean length of rest period decreases as sleep progresses, that of our first rest period being about 30 minutes and of our tenth about 8 minutes.

V. SOME TEMPERATURE FACTORS

The weather, and especially temperature, is popularly thought to have a marked influence upon the quietness of sleep. This impression is probably due to isolated experiences with conditions of extreme temperatures, such as almost anyone can recall. It is interesting, however, to study the effect of temperature on the quietness of sleep as measured by objective records.

ROOM AND OUTSIDE TEMPERATURES

In order to study the relationship between room temperature and sleep, the parents of our subjects were asked to record the room temperature each evening and morning as registered on a thermometer that was provided. Unfortunately, these readings were found to be practically worthless for our purpose, since the parents almost invariably read the thermometer in the evening just before opening the window and turning off the radiator, and in the morning after getting the room warm again. It was therefore decided to use outside temperatures. The correlations between mean rest period and room temperature are given in Table 10. Some are positive; some are negative; and only two are as much as four times their own probable errors.

The minimum temperature is recorded each day by the Minneapolis station of the United States Weather Bureau, which is three or four miles from the district where the children slept. The minimum temperature, of course, occurs at different hours on different days, though generally at night, and since it is the outside temperature and not that of the room in which the child slept, we cannot expect to find any very close relationship between temperature and quietness of sleep. Since we know, however, that the influence of outside temperature upon room temperature varies with the outside temperature, and since our records show that there was almost always a window open in the rooms where the children slept, we may reasonably believe that if any relationship exists, some indication will be found in the data.

It is a common observation that the minimum temperature for any night usually occurs after midnight. This means that the minimum temperature for any date generally occurs during the early morning of that date. The nights in this experiment, however, bear the date of the day on which they started. It may be, then, that we should correlate the mean rest period of one date with the minimum temperature of the next date; this was done for a few of the subjects,

TABLE 10.—CORRELATIONS BETWEEN LENGTH OF MRP2 AND
ROOM TEMPERATURE AT BEDTIME

Subject	No. of Nights	r	PE_r	Subject	No. of Nights	r	PE_r
1	214	−.125	.05	8	185	−.093	.05
2	53	.241	.09	9	118	−.130	.06
3	253	−.115	.04	10	272	−.098	.04
4	206	−.110	.05	11	69	−.107	.08
5	183	.120	.05	12	243	−.187	.04
6	222	.053	.05	22	112	.115	.06
7	178	−.292	.05				

TABLE 11.—CORRELATIONS BETWEEN LENGTH OF DAILY MRP2 AND MINIMUM
OUTSIDE TEMPERATURE OF THE SAME AND THE NEXT DATE FOR FOUR
SUBJECTS, ACCORDING TO SEASON

Subject	Season	MRP2 and Temperature Next Date		MRP2 and Temperature Same Date	
		r	PE_r	r	PE_r
3	Winter	−.290	.06	−.396	.06
	Summer	.076	.05	.035	.05
6	Winter	.275	.06	.213	.06
	Summer	−.130	.07	−.111	.07
10	Winter	.095	.07	.035	.07
	Summer	−.173	.06	−.172	.06
22	Winter	.366	.06	.316	.06

with results shown in Table 11. This method usually gave a closer relationship, but the difference was so small that it was not thought worth while to recalculate all the correlations by this method.

The correlations between the minimum outside temperature and the quietness of the children's sleep are shown in Table 12, along with their probable errors and the number of nights' records considered. The correlations are all small; only a few are statistically significant; and these are not all in the same direction. Consequently no conclusion can be drawn from them. The correlation based on all the records for all the subjects means very little, since in this case deviations of the variates are taken from the means of the group, while the means for the individual children differ from the means of the group by an amount that is not due to the relationship under consideration. This lack of relationship is shown by the rank-order correlation between the mean of the temperature and the mean of the rest measures on each child, which is only —.172 ± .146.

TABLE 12.—CORRELATIONS BETWEEN LENGTH OF MRP2 AND
MINIMUM OUTSIDE TEMPERATURE

Subject	No. of Nights	r	PE_r	Subject	No. of Nights	r	PE_r
1	250	−.075	.02	22	116	.316	.06
2	101	−.053	.07	13	21	.297	.12
3	280	−.168	.04	14	57	.275	.09
4	266	−.015	.04	15	70	.249	.07
5	220	−.061	.05	16	35	−.369	.11
6	267	.115	.04	18	67	.071	.08
7	259	−.212	.04	19	29	.088	.13
8	228	.032	.05	20	42	.061	.09
9	200	−.012	.05	21	66	.185	.07
10	291	−.027	.04	32	74	.331	.07
11	116	.032	.07	All . . .	3,339	−.032	.01
12	284	.044	.04				

In searching for a possible explanation of the inconsistent product-moment correlations just reported, it was found that some that were significant and positive were based on low mean temperatures, while others that were significant and negative were based on high mean temperatures. This suggests the hypothesis that when measuring the relation between temperature and quietness of sleep at low room temperatures or in cold weather, we may expect a positive relationship; when measuring it at high temperatures, a negative relationship. To test this hypothesis, the correlation coefficients just referred to were correlated with the mean temperature readings for the nights included on each child's record, the rank-order method being used; and second, the data were divided into winter and summer distributions and correlations recalculated.

In order to determine the significance of each correlation, the value of r was divided by its probable error. This ratio was given the same sign as r, making the ratio with its sign a measure of positiveness as well as of significance of the correlation. It is possible to avoid negative signs in using this method by converting the ratios into chances in a hundred or in a thousand as given in a probability table. When these ratios are correlated with the mean temperature readings for the various children, $r = -.595 \pm .10$.

TABLE 13.—CORRELATIONS BETWEEN LENGTH OF MRP2 AND MINIMUM OUTSIDE TEMPERATURE OF THE SAME DATE, ACCORDING TO SEASON

Subject	Winter				Summer			
	No. of Nights	r	PE_r	r/PE_r	No. of Nights	r	PE_r	r/PE_r
22	118	.316	.06	5.27	...	...	...	...
8	140	.225	.05	4.50	88	−.019	.07	− .27
6	127	.213	.06	3.55	140	−.111	.07	−1.59
14	57	.275	.09	3.06	...	...	...	...
13	21	.297	.12	2.48	...	...	...	...
21	32	.071	.12	0.59	34	−.142	.11	−1.29
10	150	.035	.07	0.50	141	−.172	.06	−2.87
12	147	.001	.07	0.01	137	.220	.06	3.67
15	34	−.004	.11	−0.04	36	−.226	.11	−2.05
11	67	−.023	.08	−0.29	49	.032	.10	0.32
32	31	−.065	.11	−0.59	43	−.471	.10	−4.71
5	90	−.055	.07	−0.79	130	.069	.06	−1.15
9	103	−.069	.07	−0.99	97	−.025	.07	−0.36
18	32	−.176	.11	−1.60	35	−.200	.11	−1.82
4	120	−.198	.06	−3.30	146	.146	.06	2.43
7	141	−.254	.06	−4.23	118	−.152	.06	−2.53
3	105	−.396	.06	−6.60	175	.035	.05	0.70
19	...	...	...	...	29	.088	.13	0.68
16	...	...	...	...	35	−.369	.11	−3.35
1	...	...	...	...	250	−.075	.02	−3.75
2	...	...	...	...	101	−.053	.07	−0.76
20	...	...	...	...	42	.061	.09	0.68

This correlation is significant and supports the hypothesis suggested above.

Secondly, data on each subject collected during the months of November, December, January, February, and March were correlated and the values entered under "winter." The other months are considered together and called "summer." The ratios r/PE_r were worked out, and the subjects were arranged in descending order of these ratios for the winter months. (See Table 13.) A glance at the summer and winter ratios, respectively, will show a tendency for positive correlations to obtain in winter and negative correlations in summer. This again supports our hypothesis. Moreover, keeping the seasons separate produces more significant correlations in spite of the reduced number of measures in each correlation.

These findings are not surprising if we assume that a medium temperature is best for quiet sleep. In cold weather the higher temperatures would be associated with quieter sleep, thus producing a positive correlation. The opposite would be true in summer. This reminds us of the curvilinear relationship found by Boynton and

TABLE 14.—CORRELATIONS BETWEEN TIME ASLEEP AND MINIMUM OUTSIDE TEMPERATURE OF THE SAME DATE

Subject	No. of Nights	r	PE_r	Subject	No. of Nights	r	PE_r
1	250	−.282	.04	8	228	−.168	.04
2	101	−.554	.05	9	199	.039	.05
3	279	−.410	.03	10	291	−.152	.04
4	266	−.189	.04	11	116	−.095	.06
5	220	.163	.04	12	284	.083	.04
6	266	−.322	.04	22	116	−.021	.06
7	259	−.061	.04				

Goodenough (3). They found no clear relationship between outdoor temperature and sleep, but did find a small curvilinear relationship between room temperature and length of nap, the correlation ratios being .34 and .51. Extremely low and also extremely high temperatures were unfavorable for the duration of sleep. We correlated the length of night sleep, T_2, with the minimum outside temperature, but unlike Boynton and Goodenough, we found a negative relationship and very little evidence of curvilinearity. (See Table 14.)

The relationships we found between temperature and quiet sleep are slight, partly on account of the inadequacy of the experimental conditions and of the inexact method of measuring temperature. It is not possible, therefore, on the basis of the present data to establish an optimal mean and range of temperatures for sleeping. It is suggested that the temperature of the more immediate surroundings of the subject, i. e., the bed temperature under the covers, is more important than the temperature outside or in the room.

BED AND BODY TEMPERATURES

For two of our subjects we attempted to measure bed temperature. A Tycos maximum registering thermometer was enclosed and suspended inside a half-inch tube made of wire net in such a way that the thermometer did not touch the sides of the tube. This tube was fastened with safety pins to the under side of the first blanket or sheet which covered the child and about in the center of the bed. The arrangement was such that there was always just one thickness of night clothing between the child's skin and the wire case of the thermometer, and at no time could the glass of the thermometer touch the skin or come closer to it than a quarter of an inch. Each day the thermometer was read and shaken down, then at night pinned into the bed to register the bed temperature. Each morning and evening the mother read and recorded the room temperature and the child's body temperature taken by rectum. The two children observed in this manner, a boy and a girl, were selected on the basis of the ability and willingness of their mothers to take the room and body temperature readings. Both mothers are highly intelligent and cooperative, and the writer believes the temperatures were accurately and carefully read.

For Subject 7 these observations were made daily with a few exceptions from February 13 to March 24, 1929, inclusive (body temperatures were taken till April 9); for Subject 3 they were made from March 27 to April 26, 1929. The bed temperatures for any one child were all read from the same thermometer. The children were entirely well during these periods except that for three days Subject 7 had or was suspected of having a slight cold. On these days his temperature was 100.5°, 100.2°, and 97.7° in the evenings and 99.6°, 98.8°, and 99.0°, respectively, the next mornings. Only one of these is outside the normal range for this child, as will be seen from Table 15, which gives the ranges, means, and standard deviations of the various temperatures recorded for him. The bed temperatures are interesting in comparison with Starling's figures (52, page 1222) for normal skin temperatures—33° to 35° C. or 91.4° to 95° F. The range of maximum bed temperatures for Subject 3 is from 90° to 97° and for Subject 7 from 86° to 94°.

To find the relationship between body or bed temperature and mean rest period, rank-order correlations were calculated. For Subject 3, ρ (correlation between MRP2 and evening body temperature) is —.03 ($N = 19$); for Subject 7, .04 ($N = 44$). The correlation between MRP2 and maximum bed temperature is —.04 ($N = 21$) for Subject 3, and .11 ($N = 29$) for Subject 7. These correlations are not definite enough to tell us anything, but it is possible that the lack of agreement between the subjects is a function

TABLE 15.—ROOM, BODY, AND BED TEMPERATURE READINGS FOR TWO SUBJECTS
(In Degrees Fahrenheit)

Temperature	Subject 3 (Girl)				Subject 7 (Boy)*			
	Mean	σ	No. of Readings	Range	Mean	σ	No. of Readings	Range
Room, evening .	75.0	7.6	29	70–80	71.9	1.68	50	67–78
Room, morning .	66.8	3.1	26	58–70	64.0	1.27	49	61–69
Rectal, evening .	98.8	3.2	27	98.0–100.6	99.6	.60	51	99.0–100.2
Rectal, morning .	98.5	2.7	28	97.8–100.0	99.3	.28	49	98.6–99.8
Bed, daily maximum	93.5	3.1	28	90–97	90.4	1.90	35	86–94

* The body temperature ranges as given for Subject 7 do not include the highest and lowest readings that were taken when the child was suspected of having a slight cold. All other readings taken during that period are within the normal range.

of the difference between their room temperatures. While the mean evening room temperature for Subject 7 is 71.9°, that for Subject 3 is 75.0°.

An attempt was made to trace the origin of the bed temperature as obtained. Since the metabolism of the child's body is the main source of the heat that produces this temperature, the body temperature should be considered. But the bed temperature depends not only upon the temperature of the heat source but also upon the velocity of irradiation from the bed, which in turn is influenced by the temperature of the room. It was thought possible, therefore, that some combination of these factors would be of use. For want of a theoretical basis for any more complicated method, we simply took the sum of the body and room temperatures combined. These three measures—the body temperature, the room temperature, and the combination—were each correlated with the bed temperatures. For Subject 3, the bed temperature correlated .15 ± .12 with evening room temperature, .19 ± .13 with the evening rectal temperature, and .29 ± .12 with room temperature plus rectal temperature. Similar product-moment correlations for Subject 7 were respectively .48 ± .11, —.10 ± .14, and .37 ± .12. These correlations were all based on from twenty-three to twenty-eight readings.

The relationships represented may be much closer than these correlations indicate, since the very narrow range of temperatures tends to prevent high correlations. The correlations are all positive except the smallest one, and although none of them are large, they suggest that the bed temperature is a function of the body temperature and the room temperature, and not merely of some independent factor such as the amount of bedclothes. The small number of measures and their small variability make it difficult to obtain any more definite information from correlation coefficients.

Fever temperature.—The mean rest period (MRP2) and the length of night's rest (T2) were tabulated separately for those days on which the child's body temperature was 100°, 101°, 102°, 103°, and less than 100° or unreported. The average of the child's mean rest periods for days when his temperature was less than 100° was considered the normal rest period for that child, and the average mean rest periods for the several fever temperatures were compared with this norm. The mean differences were calculated, as shown in Table 16. The same thing was done with T2, or the length of sleep. These data are summarized in the lower part of the table. The figures in the category "100° and above" include all the data presented. Since, however, some of the temperatures of 100° were reported when the child was judged to be perfectly well both by the mother and the Institute nurse, the data for 101° and above were summarized separately. Although there are only forty-two records of temperatures of 100° or higher, it is fairly clear from the agreement among them that the mean rest periods were shorter when the body temperature ranged from 100° to 103°. The differences are greater at the higher temperatures and very slight at 100°. The original data show that all the children who had temperatures of 101° or more were more restless at these temperatures, but the reverse is true of some of them at 100°. This may mean that, as the data in Table 15 indicate, 100° is still within the normal temperature range.

These findings agree with some of Karger's (31), and do not necessarily conflict with the rest, since it is quite possible that some types of fever, or temperatures above 103°, produce motionless sleep such as Karger reports. We have not been able to obtain Karger's book (29), but an unsigned abstract in the book review section of the *American Journal of Diseases of Children* (1) reports Karger as finding that "fever shows no increased activity even when produced artificially."

TABLE 16.—DEVIATIONS FROM THE NORMAL OF REST MEASURES TAKEN UNDER CONDITIONS OF HIGH BODY TEMPERATURES

Temperature	No. of Children	Total No. of Nights	Difference in MRP2	Mean Difference in Time Asleep
100°	6	28	−0.09m	−11m
101°	3	5	−1.54m	+20m
102°	3	6	−1.13m	+29m
103°	2	3	−2.78m	−44m
100° and above . . .	6	42	−0.60m	−∞.8m
101° and above . . .	4	14	−1.63m	+°2m
102° and above . . .	3	9	−1.68m	+°1m

It is also clear from Table 16 that body temperatures up to and including 103° have no marked or consistent effect upon the length of sleep. The data are not sufficiently numerous, however, to establish a negative finding.

EFFECT OF TEMPERATURE ON INITIAL ACTIVE PERIOD

The relationship between the temperature of the room at bedtime and the length of time the child tossed about before becoming quiet for at least five minutes is shown in Table 17. Most of these correlations are very low and are distributed on either side of zero. In the case of five subjects, however, the coefficients are larger. One of these is more than seven times its own probable error, one is more than five times, while the other three are only about three times their probable errors. This means that for most of the subjects, though not for all, there was no relationship between room temperature and the length of time required to become quiet after going to bed. Where the relationship did exist, we cannot be certain whether it was a peculiarity of the child or a function of external conditions. It is interesting to note that four of the five correlations mentioned are negative, and that the positive one is hardly three times its probable error. This means that in the case of those subjects for whom any correlation was found, lower room temperatures were usually associated with longer periods of initial restlessness. It may well be that for these subjects a warmer bedroom tends to induce sleep more quickly. This question should be settled by work-

TABLE 17.—CORRELATIONS BETWEEN EVENING ROOM TEMPERATURE AND LENGTH OF INITIAL ACTIVE PERIOD

Subject	No. of Nights	r	PE_r	Subject	No. of Nights	r	PE_r
1	216	.093	.05	8	187	.025	.05
2	53	−.270	.09	9	121	−.061	.06
3	261	−.237	.04	10	277	.027	.04
4	203	−.096	.05	11	82	.130	.07
5	193	−.147	.05	12	246	−.291	.04
6	232	.077	.04	22	115	.007	.06
7	205	.149	.05				

TABLE 18.—CORRELATIONS BETWEEN LENGTH OF MRP2 AND RELATIVE HUMIDITY, ACCORDING TO SEASON

Subject	Winter			Summer			Both Seasons		
	No. of Nights	r	PE_r	No. of Nights	r	PE_r	No. of Nights	r	PE_r
1	...	...	...	...	...	...	250	−.035	.04
2	...	...	...	...	...	...	101	−.335	.06
3	104	−.204	.06	176	−.117	.05	280	−.087	.04
4	120	−.252	.06	146	−.093	.06	...	...	...
5	90	−.117	.07	130	−.151	.06	...	...	...
6	127	.005	.07	140	−.013	.07	267	−.036	.04
7	141	.148	.06	118	.026	.06	...	...	...
8	140	−.019	.06	88	−.066	.07	...	...	...
9	103	−.013	.07	97	−.055	.07	...	...	...
10	188	−.197	.05	102	−.161	.06	290	−.146	.04
11	...	...	...	...	...	...	116	−.066	.06
12	146	−.232	.05	138	.190	.06	...	...	...
22	116	−.139	.06	...	...	...	...	...	...
13	21	.294	.14	...	...	...	...	...	...
14	57	−.397	.08	...	...	...	...	...	...
15	...	...	...	...	...	...	70	−.365	.07
16	...	...	...	35	.021	.11	...	...	...
18	...	...	...	...	...	...	67	−.284	.08
19	...	...	...	29	−.285	.12	...	...	...
20	...	...	...	42	.175	.10	...	...	...
21	...	...	...	...	...	...	66	−.311	.08
32	...	...	...	...	...	...	74	−.327	.07

ing with a number of children under the same conditions and at the same series of temperatures.

RELATIVE HUMIDITY

The correlations between mean rest period and relative humidity as measured at 7 P. M. by the weather bureau are so small and variable that it is difficult to state a very definite relationship, although most of the coefficients are negative and in the case of eight subjects are about four or five times their probable errors. (See Table 18.) This indicates that quieter sleep and lower humidities were associated. We hesitate, however, to make much of this concomitance, since relative humidity is not a physiological concept, and its meaning in relation to physiology is not clearly known. Consider that Minnesota winters are often characterized as "dry cold," although the physical humidity often runs above 90 per cent and sometimes as high as 100 per cent, even in January; that the summer weather is often uncomfortably humid at from 50 to 60 per cent; that daytime summer humidities seldom go above 75; and that 65 per cent is almost unbearable in hot weather. This means that the units of relative humidity do not have the same physiological significance at different temperatures, and that if physiological work is to be done with relative humidities, temperatures should be controlled.

SUMMARY

1. The room temperatures read by the mother at the child's bedtime show no consistent relationship to the quietness of his sleep.

2. Very little relationship was found between the minimum temperature outside and quietness of sleep, in spite of the fact that there was nearly always an open window in the child's bedroom.

3. What relationship was detectable tended to be positive in winter and negative in summer. That is, the extremes of temperature seemed to produce a little more activity. The correlations are small and variable, but the trends were brought out by determining the ratios of the correlations to their probable errors and by working with the seasons separately.

4. The children showed a tendency to sleep longer in cold than in warm weather, but we do not know whether the seasonal differences in length of sleep are due directly to the temperature or to other seasonal concomitants.

5. The maximum temperature under the bedclothes but outside the child's night clothes was measured nightly for a month in the case of two subjects, a boy and a girl. These temperatures approached the values for normal skin temperatures. Because of the narrow range and the small number of measures, it was not possible to determine definitely whether or not there was any relation between these bed temperatures and any of the characteristics of the subject's sleep. Bed temperatures do, however, seem to be related to the body temperature and also to room temperature, and thus are not entirely a function of some independent factor such as the amount of bedclothes.

6. Fever temperatures of 100° to 103° were associated with less quiet sleep, but the difference was so slight as to be almost negligible.

7. Some children went to sleep more quickly when their room was warmer, but most of them showed no consistent difference. We do not know whether those who go to sleep quickly in warm rooms are different from the others, or whether the something different is in the home conditions.

8. The children tended to sleep more quietly when there were lower relative humidities, but since temperature was not controlled, we do not know the significance of this finding.

VI. AGE, SEASONAL, AND SEX DIFFERENCES

A number of recent studies have been concerned with the relation of length of sleep to age, seasons of the year, and sex (2, 7, 16). We have studied the relation of these factors to quietness of sleep as well as to length of sleep, and while we have not worked with the large numbers of children involved in some of the recent surveys, we do have a considerable number of records on each child within each seasonal, age, and sex group.

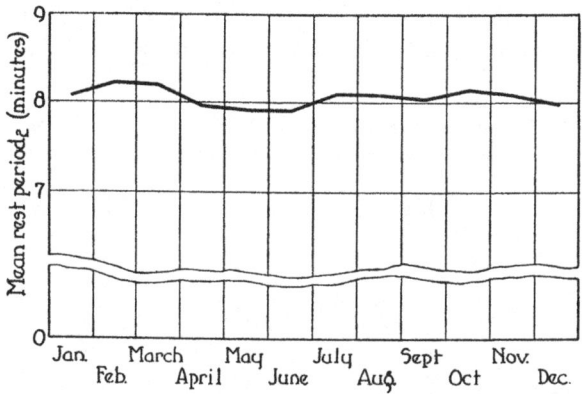

FIGURE 11.—MEAN REST PERIODS OF THE CHILDREN, ACCORDING TO MONTH OF YEAR

DIFFERENCES IN MEAN REST PERIODS

In Table 19 and Figure 11 the mean rest periods of the entire group of children are presented for each month of the year. From the curve in Figure 11 it appears that the children were more active

TABLE 19.—LENGTH OF MRP2 DURING DIFFERENT MONTHS

Month	No. of Nights	MRP2	σ	Month	No. of Nights	MRP2	σ
Jan.	236	8.08m	0.90	July	186	8.07m	1.49
Feb.	307	8.21m	1.44	Aug.	162	8.07m	1.65
March	422	8.22m	1.43	Sept.	49	8.02m	1.83
April	415	7.96m	1.38	Oct.	206	8.13m	1.51
May	427	7.89m	1.32	Nov.	282	8.08m	1.51
June	266	7.90m	1.28	Dec..	307	7.97m	1.47

in the summer months (April through October) and most quiet in February and March. Table 20 shows that some of these differences are statistically significant.

When the mean rest periods are summarized according to age instead of season, some rather large differences appear, as will be seen in Table 21 and Figure 12. Taken together, these two facts mean that the obtained differences in both cases are probably due to a combination of the factors of season and age, and we cannot tell to what extent each is a function of the other. To label a curve an age curve does not make it so. Its fluctuations may be due to seasonal influence unless we rule out the latter in some way. The method used here is simple in principle, and though it seems not to be in very general use, is very effective where sufficient data are available.

TABLE 20.—DIFFERENCES BETWEEN LENGTH OF MRP2 IN VARIOUS SEASONS

Season	MRP2	σ*	Diff.	$\dfrac{\text{Diff.}}{\sigma \text{ Diff.}}$
Feb.–March	8.21m	1.47	− .29	−4.33
April–June.	7.92m	1.34	− .14	−1.56
July–Sept.	8.06m	1.62	− .04	−0.04
Oct.–Nov.	8.10m	1.51	.08	0.09
Dec.–Jan.	8.02m	1.24	− .19	−2.50
Feb.–March† . . .	8.21m	1.47	.29	4.33
April–June	7.92m	1.34	− .14	−2.55
July–Jan.	8.06m	1.44	− .15	−2.27

* The sigmas are those of the distributions of mean rest periods, not of the means of those distributions.

† After this point the months are rearranged into new groups between which the differences are more significant.

TABLE 21.—LENGTH OF MRP2 ACCORDING TO AGE OF CHILD

Age (in months)	No. of Nights	MRP2	σ	Age (in months)	No. of Nights	MRP2	σ
24–26. . . .	53	9.04m	1.81	42–44. . . .	479	7.92m	1.53
27–29. . . .	209	8.15m	1.61	45–47. . . .	408	8.48m	1.22
30–32. . . .	209	8.14m	1.44	48–50. . . .	289	8.43m	1.25
33–35. . . .	355	8.07m	1.50	51–53. . . .	243	8.35m	1.27
36–38. . . .	388	7.84m	1.41	54–56. . . .	135	7.96m	1.17
39–41. . . .	455	7.34m	1.30	57–59. . . .	42	8.25m	1.25

It is hereby recommended for problems of this nature. The essential feature of this method consists in the simultaneous separation of the data into age and season categories: i. e., the measures at a given age were summarized separately for the different seasons, and the data for each age separately were summarized for each season separately. When the data are arranged in a coordinate table such as Table 22, an age curve can be made for each season by plotting the horizontal columns of values, as in Figure 13, and a seasonal curve for each age by plotting the vertical columns of values, as in Figure 14. These curves show the effect of one factor divorced

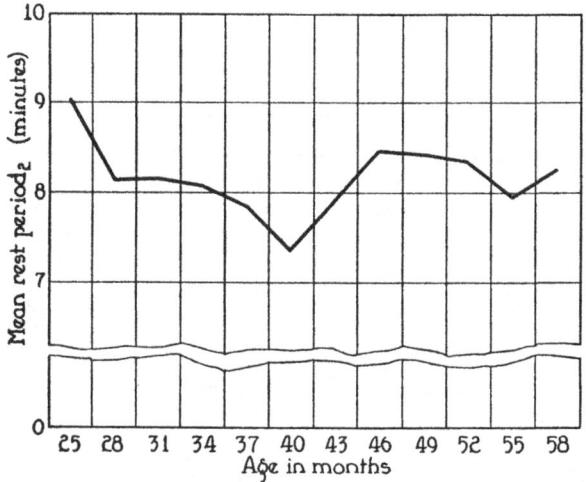

FIGURE 12.—MEAN REST PERIODS OF THE CHILDREN,
ACCORDING TO AGE

from the influence of the other. We can see now that the age factor produces much larger and more consistent differences than does the seasonal factor.

If we wish to combine the age curves of the various seasons into a single age curve, we may do so by finding the amount by which each point on the age curve for a given season deviates from the mean for that season and adding this deviation algebraically to the deviations of the corresponding points on the other age curves. The result of this procedure is shown in the lower part of Figure 13, where the points plotted are the means of the deviations described, weighted by the number of nightly records in each age-season cate-

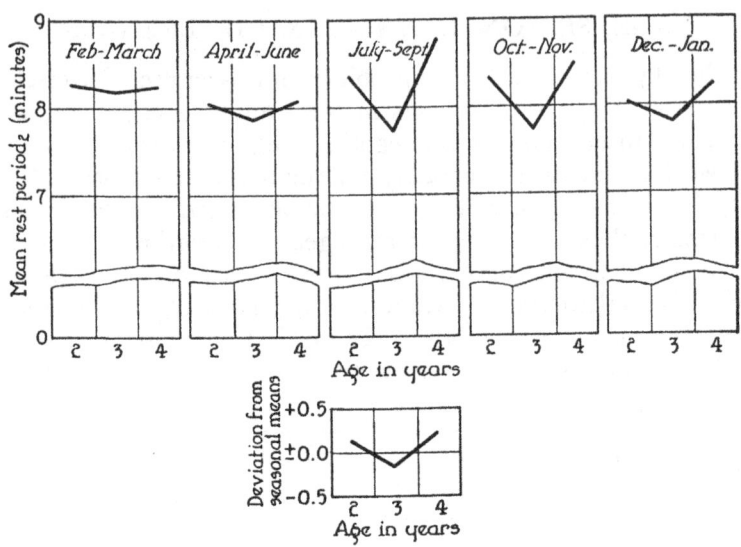

FIGURE 13.—MEAN REST PERIODS OF CHILDREN OF VARIOUS AGES,
ACCORDING TO TIME OF YEAR

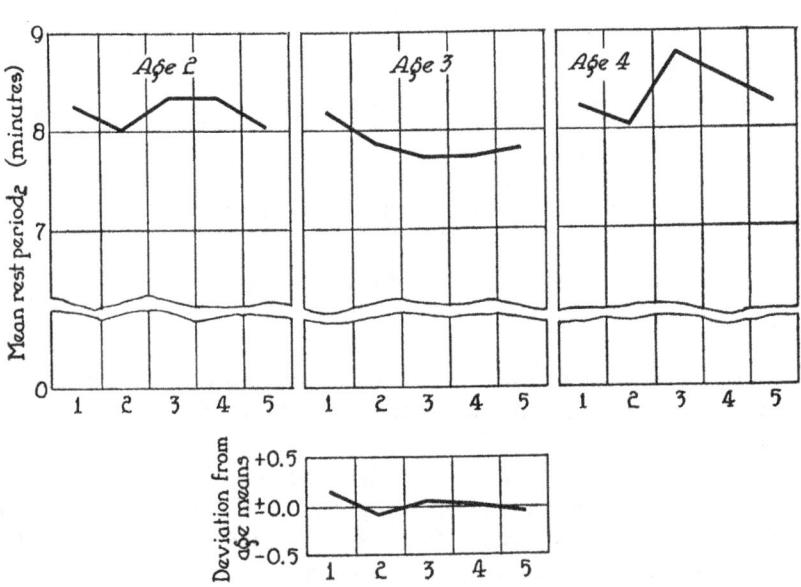

Seasons: 1, Feb.-March; 2, April-June; 3, July-Sept.; 4, Oct.-Nov.; 5, Dec.- Jan.

FIGURE 14.—MEAN REST PERIODS OF CHILDREN DURING VARIOUS MONTHS,
ACCORDING TO AGE

gory. Here again the age difference stands out. When the seasonal curves of Figure 14, which are not consistent at the different ages, are thus combined in the lower part of the figure, the resulting seasonal differences look very small. The only ones of considerable size are the increase in mean rest period from December and January to February and March, and the drop from February and

TABLE 22.—LENGTH OF MRP2 ACCORDING TO AGE OF CHILD AND TO SEASON

Season	24–35 Months			36–47 Months			48–59 Months		
	No. of Nights	MRP2	σ	No. of Nights	MRP2	σ	No. of Nights	MRP2	σ
Feb.–March .	174	8.26m	1.89	378	8.17m	1.43	177	8.22m	1.19
April–June . .	349	8.01m	1.54	661	7.85m	1.25	98	8.04m	1.12
July–Sept. . .	140	8.34m	1.64	207	7.71m	1.56	50	8.76m	1.20
Oct.–Nov. . .	104	8.33m	1.41	227	7.73m	1.61	157	8.49m	1.24
Dec.–Jan. . .	59	8.03m	1.08	257	7.81m	1.55	227	8.25m	1.33

March to April, May, and June. To summarize, we find a tendency for children to sleep more quietly in February and March, a slight tendency for them to be more restless in April, May, and June, and a very slight tendency toward restlessness in December and January. With respect to age differences, we find the three-year-old children significantly more active in their sleep than the two-year olds or the four-year-olds. The four-year-old children were the least active in the group.

SEX DIFFERENCES

The fact that the three-year-old children were the most active is a little surprising, especially if we suppose that a change with age should be continuous. Since it was suspected that the finding might have been affected in part by a possible sex difference and unequal proportions of sex data at the various ages, the differences between the sexes were worked out. A glance at Table 23 will show that the sex differences in the various rest measures are very small, but some of them are statistically significant. Without large numbers of cases it would be impossible to determine the direction of such small differences. The girls stayed in bed eight minutes longer than the boys, a difference of only 1.27 per cent, but this difference is 3.4 times its standard error, or in other words the chances are better than 999 in 1,000 that it is not due to errors of random sampling. The differences are all of the order of 1 per cent except in the case of the initial active period, which shows that the girls require more than two minutes longer to become quiet after retiring, or nearly 7 per cent longer than the boys.

The mean rest period for boys is only 0.09 minutes longer than that for the girls; consequently there would have to be an undue proportion of girls' records at age three in order even to help

TABLE 23.—SEX DIFFERENCES IN SEVERAL REST MEASURES*

Measure	Girls	Boys	Diff.	Diff. / σ Diff.
Time in bed	11h 11m	11h 03m	8.00m	3.4
Time asleep	10h 24m	10h 19m	5.00m	2.3
MRP1	7.22m	7.35m	0.13m	2.6
MRP2	7.99m	8.08m	0.09m	1.6
Initial active period . . .	34.90m	32.65m	2.25m	2.3

* In all measures except the initial active period, the product of the number of observations and the number of girls is 1,175 and of number of observations and number of boys is 2,090. For the initial active period the product for girls is 1,224 and for boys 2,188. Data on Subject 32 are not included.

produce our apparent age difference. Actually there is not; the percentage proportion of girls' records to boys' at age two is 54 to 46; at age three, 36 to 64; and at age four, 11 to 89. The disproportion in the sex data at age three, even if it were effective, is in the wrong direction to produce the finding suggested. If these sex proportions were effective, age two would be the most active, which it clearly is not. The fact remains, then, that whatever the explanation, the three-year-old children are the most active sleepers of the group.

DIFFERENCES IN LENGTH OF SLEEP

When the length of sleep, T2, is summarized according to age and season (see Table 24 and Figures 15 and 16) we find only a very slight tendency for shorter sleep in summer and for shorter sleep with increasing age. Again, however, the two factors must be separated, as was done in the case of the mean rest period. (See Table 25 and Figures 17 and 18.) Here the seasonal differences persist and become more pronounced at the expense of the apparent age differences. Now we can see (Figure 18) a consistent tendency at all three ages for sleep to be longer in winter and shorter in summer. This agrees with the finding of the survey undertaken by the Institute of Child Welfare of the University of Minnesota

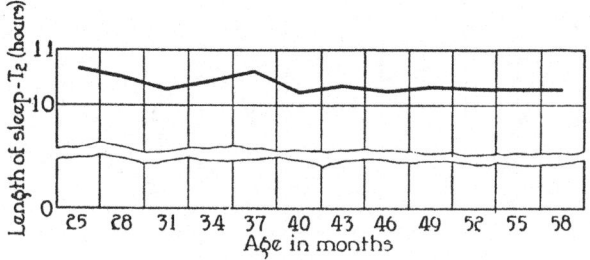

FIGURE 15.—MEAN LENGTH OF CHILDREN'S SLEEP IN
HOURS, ACCORDING TO AGE

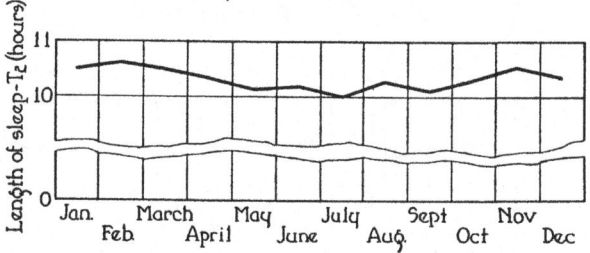

FIGURE 16.—MEAN LENGTH OF CHILDREN'S SLEEP IN
HOURS, ACCORDING TO MONTH OF YEAR

TABLE 24.—MEAN TIME ASLEEP (T2) ACCORDING TO AGE OF
CHILD AND MONTH OF YEAR

Age (in months)	No. of Nights	T2	σ	Month	No. of Nights	T2	σ
24–26 . . .	53	10h 39m	36.0m	Jan. . . .	236	10h 29m	87.0m
27–29 . . .	209	10h 29m	62.5m	Feb. . . .	307	10h 37m	56.5m
30–32 . . .	209	10h 16m	55.0m	March . .	422	10h 30m	67.0m
33–35 . . .	355	10h 24m	63.0m	April . . .	415	10h 21m	56.0m
36–38 . . .	388	10h 36m	60.0m	May . . .	427	10h 08m	59.0m
39–41 . . .	455	10h 13m	59.5m	June . . .	266	10h 10m	66.0m
42–44 . . .	479	10h 21m	58.0m	July . . .	186	10h 01m	55.0m
45–47 . . .	408	10h 13m	64.5m	Aug. . . .	162	10h 14m	63.5m
48–50 . . .	289	10h 17m	65.0m	Sept. . . .	49	10h 06m	62.0m
51–53 . . .	243	10h 16m	70.5m	Oct. . . .	206	10h 17m	59.0m
54–56 . . .	135	10h 15m	77.0m	Nov. . . .	282	10h 29m	60.5m
57–59 . . .	42	10h 15m	62.0m	Dec.. . . .	307	10h 21m	63.0m

(2) and with Hayashi's study of Japanese school children (16). We also see that the age differences are not consistent throughout the various seasons, but that the apparent decrease in length of sleep with increase in age (Figure 15) is due entirely to such a decrease

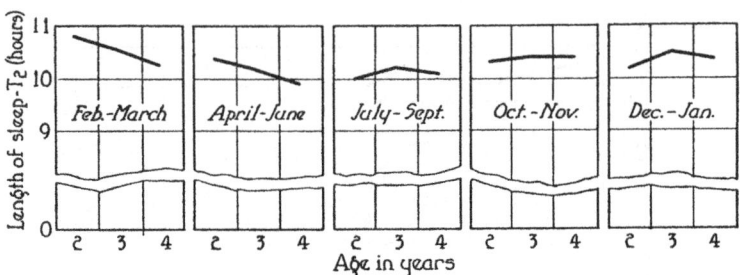

FIGURE 17.—MEAN LENGTH OF SLEEP OF CHILDREN OF VARIOUS AGES, ACCORDING TO MONTH OF YEAR

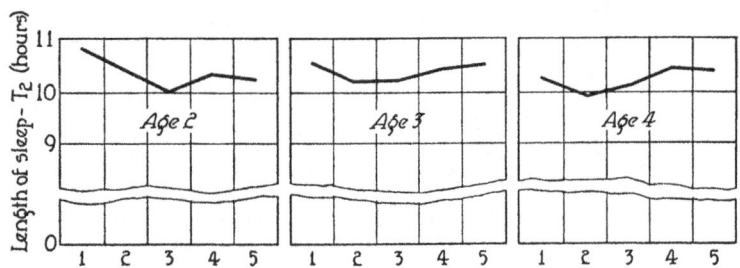

Seasons: 1, Feb.-March; 2, April-June; 3, July-Sept.; 4, Oct.-Nov.; 5, Dec.-Jan.

FIGURE 18.—MEAN LENGTH OF CHILDREN'S SLEEP DURING VARIOUS SEASONS, ACCORDING TO AGE

TABLE 25.—MEAN TIME ASLEEP (T2) ACCORDING TO AGE OF CHILD AND SEASON

Season	24–35 Months			36–47 Months			48–59 Months		
	No. of Nights	T2	σ	No. of Nights	T2	σ	No. of Nights	T2	σ
Feb.–Mar.	174	10h 50m	59.5m	378	10h 33m	60.0m	177	10h 16m	72.5m
April–June	349	10h 23m	60.0m	661	10h 11m	59.0m	98	9h 53m	54.5m
July–Sept.	140	10h 01m	55.5m	207	10h 12m	57.0m	50	10h 05m	68.5m
Oct.–Nov.	104	10h 20m	68.0m	227	10h 25m	46.0m	157	10h 26m	65.6m
Dec.–Jan.	59	10h 12m	56.5m	257	10h 30m	62.5m	227	10h 22m	70.5m

from February through June, in spite of a somewhat indefinite or even an opposite trend in the remaining seven months of the year. Foster, Goodenough, and Anderson (7) likewise found very little decrease in length of night sleep with advancing age within the age groups studied.

DIFFERENCES IN LENGTH OF NIGHT

The length of the child's stay in bed, T1, was analyzed (see Tables 26 and 27) by seasons and ages according to the above plan, but the curves are not reproduced here. Very little can be told from the age curve without controlling season; and when season is controlled no consistent age differences are obtained. When age is controlled, the seasonal differences resemble those for the length of sleep.

TABLE 26.—MEAN TIME IN BED (T1) ACCORDING TO AGE OF CHILD

Age (in months)	No. of Nights	T1	σ	Age (in months)	No. of Nights	T1	σ
24–26. . . .	53	11h 07m	36.5m	42–44. . . .	479	11h 08m	65.0m
27–29. . . .	209	11h 05m	70.5m	45–47. . . .	408	10h 56m	62.0m
30–32. . . .	209	10h 49m	68.5m	48–50. . . .	289	11h 08m	74.5m
33–35. . . .	355	11h 12m	66.0m	51–53. . . .	243	11h 09m	85.0m
36–38. . . .	388	11h 19m	62.5m	54–56. . . .	135	11h 01m	88.5m
39–41. . . .	455	11h 06m	57.5m	57–59. . . .	42	10h 50m	59.0m

TABLE 27.—MEAN TIME IN BED (T1) ACCORDING TO AGE OF CHILD AND SEASON

Season	24–35 Months			36–47 Months			48–59 Months		
	No. of Nights	T1	σ	No. of Nights	T1	σ	No. of Nights	T1	σ
Feb.–Mar.	174	11h 24m	46.0m	378	11h 17m	63.5m	177	11h 02m	91.0m
April–June	349	11h 07m	65.0m	661	11h 00m	70.0m	98	10h 33m	59.5m
July–Sept.	140	10h 37m	79.0m	207	10h 57m	62.0m	50	10h 49m	67.5m
Oct.–Nov.	104	11h 12m	71.0m	227	11h 05m	74.5m	157	11h 21m	80.5m
Dec.–Jan.	59	10h 44m	66.0m	257	11h 18m	63.0m	227	11h 16m	82.5m

DIFFERENCES IN INITIAL ACTIVE PERIOD

When the initial active period is summarized (see Table 28) and the results plotted, the data show some rather striking trends. The influence of age and that of season are separated in Table 29. It is clear that in all seasons except October–November three-year-old children had the longest average initial active period. This longer initial active period at age three seems to be corroborated by the findings of some other authors, who apparently considered it a re-

TABLE 28.—MEAN LENGTH OF INITIAL ACTIVE PERIOD DURING DIFFERENT MONTHS AND AT DIFFERENT AGES

Month	No. of Nights	Initial Active Period	σ	Age (in months)	No. of Nights	Initial Active Period	σ
Jan.	246	36.4m	29.2	24–26	58	23.8m	18.6
Feb.	327	30.1m	26.1	27–29	215	28.2m	24.1
March. . . .	444	29.7m	27.0	30–32	212	27.0m	23.2
April	434	34.4m	28.4	33–35	363	34.5m	29.8
May	437	36.7m	27.6	36–38	394	33.2m	26.4
June	276	37.3m	29.2	39–41	472	41.2m	29.0
July	193	31.7m	27.2	42–44	518	35.4m	26.7
Aug.	165	29.9m	25.3	45–47	438	30.3m	27.8
Sept.	48	28.4m	25.4	48–50	307	34.5m	28.1
Oct.	217	28.7m	21.0	51–53	249	37.2m	33.3
Nov.	297	35.2m	30.2	54–56	143	25.2m	29.5
Dec..	328	36.4m	32.2	57–59	46	23.8m	16.3

TABLE 29.—MEAN LENGTH OF INITIAL ACTIVE PERIOD ACCORDING TO AGE OF CHILD AND SEASON

Season	24–35 Months			36–47 Months			48–59 Months		
	No. of Nights	Initial Active Period	σ	No. of Nights	Initial Active Period	σ	No. of Nights	Initial Active Period	σ
Feb.–March	188	27.2m	19.9	398	35.8m	29.2	185	26.5m	26.0
April–June	354	34.3m	26.4	589	37.6m	30.5	100	35.0m	24.7
July–Sept.	140	27.5m	25.4	213	32.2m	25.0	53	32.1m	32.2
Oct.–Nov.	106	31.3m	37.6	242	32.5m	19.7	166	33.1m	27.7
Dec.–Jan.	60	20.8m	18.2	273	39.1m	29.1	241	37.1m	34.8

sult of chance fluctuations or errors of sampling. Boynton found the mean time in bed before going to sleep to be 29, 35, and 32 minutes for ages two, three, and four, respectively (8, page 148). The Institute of Child Welfare of the University of Minnesota found the mean time before sleep to be 18, 24, and 23 minutes for ages two, three, and four, respectively, a small difference, but based on a large number of cases (8, page 147). Neither Boynton nor the authors of the Institute study published their figures, but they say that no consistent age differences appear (3, page 273, and 2, page 10).

Our finding in regard to seasonal effects recalls the similar age difference in nocturnal motility as measured by the mean rest period. Can it be that age three is a restless age? At this age the children studied take longer to go to sleep and are more active during their sleep than they are at either age two or age four.

Diurnal Motility

It would be interesting to see data on diurnal motility for these ages separately. Challman (13, pages 34–37) has data on diurnal activity over virtually this same age range, but his age groupings (24–45 and 45–57 months) are such that they would obscure any fluctuation of the kind found in the present study. One cannot be confident of finding agreement between nocturnal and diurnal activity, especially in view of some comparisons that we have made in cooperation with Challman. Since he had made observations on some of the children studied here, we calculated rank-order correlations between the nocturnal and diurnal activity ratings of the two groups. The correlation between Challman's diurnal and our nocturnal activity ratings on eleven children is only .11 ± .17; on seventeen children (including the above eleven) it is .29 ± .14. This would tend to indicate that activity as a trait is not very well generalized, but instead is specific to the general type of activity studied, i. e., nocturnal or diurnal. Likewise the rank-order correlation between diurnal activity and mean length of initial active period in bed is only .02 ($N = 17$), and between initial active period and nocturnal activity, .11 ($N = 17$). This does not necessarily mean, however, that age differences in diurnal and nocturnal activity cannot show similar trends, and we should like to see the comparison made.

Summary

1. Age and seasonal differences were so combined in this study that we could not tell which factor was the effective one in producing a difference in activity unless we controlled the other factor. The factor of age was therefore studied with season held constant by working with each season separately; the effect of season was studied by controlling age in the same way. A composite curve showing the effect of age was constructed by averaging the amounts

by which each age deviated from the mean for all children at each season. Similarly, a composite curve showing the effect of season was made by taking deviation of the age-season values from the means for the several ages. It was thus found that the age factor produced differences in the mean rest period larger and more consistent than those produced by the seasonal factors.

2. Our data showed children of three years to be more active during sleep than those of either two or four, the latter being the quietest of the three groups.

3. The children's sleep in February and March was a little more quiet than the average for the year, and in April, May, and June it was a very little more restless than the average.

4. There was a consistent tendency at all three ages for sleep to be longer in winter and shorter in summer. The differences were small, however, the average for the several seasons being between ten and ten and one-half hours.

5. There was very little change in length of sleep from age two to age four. Although the older children slept considerably less than the younger ones in the spring, there were some reversals from July to January, and the averages are almost the same.

6. The length of time spent in bed showed no relationship to age when season was controlled.

7. When age was held constant, the small seasonal differences were similar to those in the length of sleep, being longer in winter and shorter in summer. The average time in bed was about eleven hours.

8. At age three the children took longer to go to sleep than at either age two or age four. This finding, together with conclusion 2, can be interpreted to mean that three-year-old children move about more in bed than others, both before and after they go to sleep. The question is raised whether they also take more exercise in the daytime. The matter of comparative activity at different ages could also be worked out with daytime naps.

9. Our children went to sleep most readily in February and March, and least readily in April, May, and June. During the other months of the year they varied but little from the average.

10. The sex differences are all small, but some of them may be slightly significant. The girls were kept in bed an average of eight minutes longer than the boys, but they slept only five minutes longer. The boys rested a little more quietly over the whole night

than the girls. They went to sleep more promptly than the girls and slept slightly more quietly. There are large individual differences, however, which keep these differences from being very significant statistically.

11. The sex differences combined with unequal numbers of boys' and girls' records at the three ages could not have produced the age differences mentioned in conclusions 2 and 8, because the differences are small and the inequality of the sex ratios is in the wrong direction.

12. The fact that the girls were in bed an average of eight minutes longer than the boys but slept only five minutes longer suggests that the length of stay in bed was regulated by the parents rather than by the child and his need for sleep. This emphasizes again the necessity for determining standards of child treatment by observation of a child's own spontaneous responses to situations. A survey conducted by the Institute of Child Welfare at the University of Minnesota (2), some of the results of which have been recorded by Foster, Goodenough, and Anderson (7), shows that children do not sleep so long as the specialists and the practical books say they should. The conviction is growing that the child's behavior is a better guide than recommendations of specialists based on impressions rather than on careful research. The same principle is illustrated in more striking fashion by the work of Davis in allowing infants to choose their own diet (5, 6).

VII. MISCELLANEOUS PROBLEMS

EFFECT OF NURSERY SCHOOL ATTENDANCE

Our data were classified according to the child's presence in or absence from nursery school on the date of each record. The days on which the child was absent included Saturdays, Sundays, holidays, days on which he or some member of his family was ill or suspected of having been exposed to some disease, and nursery school vacations, such as parts of August and September. There seem to be individual differences in the effect of attendance, or of the factors involved in absence, upon the quietness of sleep. Some children slept more quietly on nights following attendance at the nursery school than on other nights, whereas other children did just the reverse. (See Table 30.) Of the differences that are larger than their own standard errors, three show quieter sleep associated with attendance and four show the opposite. If the routine of attendance at the nursery school had no constant directional effect upon the quietness of sleep, but had a regulatory influence, we might expect to find the difference revealed in the variability of the measures rather than in the averages. Even here, however, we find very little difference. In six cases the sigmas are larger for the records taken during the child's absence from school, in three they are larger for days when he was present, and in four cases there is practically no difference. If the coefficients of variability are considered instead of the sigmas, these cases number eight, four, and one, respectively. This is very meager evidence of a regulatory influence exerted by nursery school routine.

With respect to the length of sleep, we find that of those children who showed differences larger than the sigma of the differences, three slept longer on nights following nursery school attendance and seven not so long. Since two of the former differences are four and six times their sigmas, respectively, we must conclude that in this respect, too, there were significant individual differences, although the majority of the group were alike in sleeping shorter hours if they had been at school during the day. Comparison of the sigmas of the distributions show that for nine of the thirteen children the length of sleep was more variable when the child was not

68

attending school; for the four others, less variable. The corresponding figures for the coefficient of variability instead of the sigma are eight and five. The length of sleep tended to be shorter but more regular when the child attended school.

In regard to the initial active period or length of time before sleep, eight of the thirteen children showed differences larger than the sigmas of the differences. Of these, six required more and two

TABLE 30.—DIFFERENCE AND SIGNIFICANCE OF DIFFERENCE BETWEEN REST MEASURES RECORDED WHEN CHILD ATTENDED AND DID NOT ATTEND NURSERY SCHOOL
(In Minutes)

Subject	Difference in MRP2	Difference Ratios for Time Asleep	Differences in Initial Active Period	Subject	Difference in MRP2	Difference Ratios for Time Asleep	Differences in Initial Active Period
1. . . .	5.74*	5.98*	2.52	8 . . .	0.05*	1.11*	2.07
2. . . .	0.32	0.25	2.52*	9 . . .	1.06*	1.23	2.97*
3. . . .	1.33*	4.06*	0.09	10 . . .	1.86	2.45	0.81*
4. . . .	0.85	0.91*	0.03	11 . . .	1.81	1.02	1.46*
5. . . .	2.47	0.13*	0.97*	12 . . .	2.92	3.74	6.49*
6. . . .	0.02	2.66	3.99*	22 . . .	0.02	2.11	0.29*
7. . . .	0.19*	2.50	4.09*				

* Asterisks after numbers indicate that measures were larger when the child attended nursery school.

required less time to go to sleep on the evenings following nursery school attendance. Seven sigmas are larger for the evenings following attendance and six are smaller. Only three coefficients of variability are larger for the evenings following attendance, however, while ten are smaller. In spite of individual differences, then, there seems to be a group tendency to be slower about going to sleep on nights following the days in school, and the variability as measured by the sigmas shows no consistent trend. The relative variability, however, indicates that the time required to go to sleep was more constant following days in the nursery school.

EFFECT OF ENURESIS

The parents' reports on enuresis were classified in all cases where there was a sufficient number for statistical treatment, and the mean rest period was calculated for those nights on which the bed was reported wet, for all other nights, and for those nights on which the bed was reported dry. The second category includes the third, but this is permissible since parents are more likely to omit negative than positive information, as is shown by the fact that, except in the case of Subjects 2, 7, and 12, the mean for the nights not reported is nearer to the mean for the dry than to that for the wet nights. The result of the classifiaction is given in Table 31.

The significances (diff./σ diff.) of the differences for nights on which the bed was reported wet, minus all other nights, were as follows: for Subject 22, 2.40; 7, 2.40; 12, —1.97; 3, 0.98; 2, 0.78; 10, 0.52; 15, 0.48; 8, 0.48; 4, 0.48; 9, 0.45. For nights when the bed was reported wet, minus nights when it was reported dry, the significances were, in the same order, 0.60, —0.36, —2.72, 0.95,

TABLE 31.—LENGTH OF MRP2 ON NIGHTS WHEN BED WAS AND WAS NOT REPORTED WET

Subject	Bed Reported Wet			All Other Nights			Bed Reported Dry		
	No. of Nights	MRP2	σ	No. of Nights	MRP2	σ	No. of Nights	MRP2	σ
22	19	10.00m	1.30	97	9.22m	1.28	29	9.77m	1.29
7	171	8.53m	1.21	88	8.17m	1.11	28	8.60m	0.89
12	151	8.11m	1.10	133	8.36m	1.15	71	8.54m	1.10
3	148	8.74m	1.79	132	8.55m	1.43	93	8.53m	1.59
2	8	9.54m	1.65	93	9.07m	1.52	82	8.97m	1.37
10	3	7.58m	0.66	288	7.38m	1.09	262	7.39m	0.90
15	12	6.91m	1.12	58	7.08m	1.13	43	7.03m	1.02
8	14	6.79m	1.47	214	6.98m	0.84	147	6.84m	0.89
4	9	8.32m	0.97	257	8.48m	1.50	186	8.37m	2.51
9	14	8.72m	0.84	186	8.83m	1.37	98	8.85m	1.31

0.95, 0.49, —0.33, —0.13, —0.13, and —0.50. For Subjects 22, 3, 2, and 10 the mean rest period was greater on nights when the bed was reported wet than on all other nights, and for Subject 7 this period was greater on nights when it was reported wet than on nights when it was reported dry. With one exception, the more significant differences are in favor of quieter sleep on nights when the bed was wet. (The notable exception is Subject 12, a child whose habit training had been neglected and who, up to the time of his entrance into the university nursery school, had not even established bowel control. Since this defect was quickly remedied by the nursery school teachers, we know that it was due to lack of training.)

A tendency is apparent in this group of children for enuresis to be accompanied by more quiet sleep. A wealth of interesting speculation could be written to try to explain this finding. Suffice it to say, however, that a source of disturbing irritation, which probably would otherwise persist and increase the child's motility, is removed by emptying the bladder, and that the deeper the sleep the more likely is the bladder stimulation to rise to the point of releasing the sphincter mechanism without waking the child.

Effect of Violent Exercise

The parents' reports on the activities of the child between supper and bedtime were used to separate the data into two groups on the basis of the amount of exercise taken by the child. The mean rest

Table 32.—Length of MRP2 on Nights Following Violent Exercise and on Other Nights

Subject	Violent Exercise Reported			All Other Nights		
	No. of Nights	MRP2	σ	No. of Nights	MRP2	σ
3.	14	9.24m	1.77	266	8.62m	1.62
4.	4	8.31m	0.96	262	8.48m	1.48
5.	71	8.20m	1.17	149	8.18m	1.13
7.	1	9.05m	. . .	258	8.40m	1.23
10.	118	7.39m	1.09	173	7.38m	1.07
11.	1	10.63m	. . .	115	8.93m	1.43
12.	12	8.61m	1.31	272	8.21m	1.12
22.	23	9.11m	1.55	93	9.41m	1.22

period was calculated for all those days on which violent exercise was reported and for all those on which it was not reported. This latter category includes quiet play and a small number of days on which no report was made, usually because the parents considered that there was nothing to report.

The number of records of violent exercise was usually so small for each child that no very adequate statistical treatment is possible. Table 32 shows, however, that the differences in mean rest period are small between nights following violent exercise and other nights, and do not all point in the same direction. Six of the eight differences do point toward a more quiet night after violent play, but some of these differences are extremely small. It is interesting to note that the largest differences were obtained where only one or a very few reports of violent exercise were available. The most significant differences occur in the case of Subjects 3, 12, and 22; the differences are respectively 1.28, 1.04, and 0.86 times their own standard deviations. The first two of these are in the same direction as the majority of the differences mentioned above, but the third indicates more active sleep following violent exercise in the evening. The only conclusion that can be drawn, even tentatively, is that if larger numbers of active and quiet evenings were studied, one might expect to find that physical exercise in the evening tends to produce more quiet sleep. The reverse might be expected, however, if the exercise was too violent or too long continued.

Mönninghoff and Piesbergen (42) report that they were more sensitive to experimental noise on nights following unusual exercise such as long walks; but since they drank as many as four glasses of beer before bedtime, it is possible that their finding is complicated by reinforcing stimuli from the quantity of liquid in the stomach and bladder.

Effect of Various Bowel Conditions

The length of night sleep and the mean rest period, respectively, on nights when the parents reported looseness, constipation, or normal bowel condition are recorded in Table 33. By taking devia-

TABLE 33.—LENGTH OF MRP2 AND MEAN TIME ASLEEP (T2)
DURING VARIOUS CONDITIONS OF THE BOWELS

Sub-ject*	Loose			Constipated			Normal		
	MRP2	T2	No. of Nights	MRP2	T2	No. of Nights	MRP2	T2	No. of Nights
1. . .	6.35m	11h 28m	1	5.27m	10h 18m	9	6.45m	10h 23m	229
2. . .			0	10.01m	10h 15m	2	9.03m	10h 33m	86
3. . .	6.70m	10h 33m	1	8.52m	10h 13m	87	8.72m	10h 07m	175
4. . .	7.44m	9h 55m	4	8.16m	9h 23m	1	8.41m	10h 17m	211
5. . .	8.28m	9h 48m	17	9.89m	9h 28m	1	8.19m	9h 28m	182
6. . .			0	7.26m	11h 25m	2	8.30m	11h 03m	179
7. . .	5.93m	7h 33m	1	8.20m	10h 31m	7	8.57m	10h 19m	189
8. . .	7.41m	11h 13m	3	6.91m	10h 16m	16	6.84m	10h 31m	157
10. . .	7.73m	11h 12m	17	7.48m	11h 05m	29	7.34m	10h 55m	226
11. . .			0	7.49m	9h 40m	2	8.85m	9h 37m	88
12. . .	7.66m	10h 18m	1	10.40m	10h 43m	1	8.16m	10h 06m	105
22. . .	8.83m	9h 13m	1	9.58m	10h 46m	3	9.31m	9h 59m	103

* The mother's reports on this section for Subject 9 were all blank.

tions of the means of the special bowel conditions from the normal means for each child, we determined the general tendency of the differences. We found no consistent difference in the quietness of sleep between normal nights and nights when the child was constipated, although this may have been due to lack of a sufficient number of records. On the other hand, we found a tendency toward greater restlessness when the bowels were loose than when they were normal. With respect to the length of sleep, the differences were not very consistent, but there did appear to be a slight tendency to sleep longer when constipated than when normal. Since there may be individual differences, this finding should be verified by more records on a larger number of children. With respect to the comparison between looseness and normal bowel condition, the algebraic sum of the deviations would seem to indicate that there is also a slight tendency to sleep longer when the bowels are loose than when normal, but this is due to large differences in one or two cases, and the reverse is true of most of the subjects.

Effect of Bedtime on Length of Stay in Bed

Correlations between time of going to bed and length of time in bed for thirteen subjects were as follows: 1, $-327 \pm .04$; 2, $-.478 \pm .05$; 3, $-.442 \pm .03$; 4, $-.710 \pm .02$; 5, $-205 \pm .04$; 6, $-.449 \pm .03$; 7, $-.292 \pm .04$; 8, $-.435 \pm .04$; 9, $-.614 \pm .03$; 10, $-.664 \pm .02$; 11, $-.236 \pm .06$; 12, $-.532 \pm .03$; 22, $-.506 \pm .05$. These negative correlations indicate that there is a tendency to arise at a somewhat constant time in the morning regardless of the time of going to bed. If the child *had a certain amount of sleep* regardless of the time of going to bed, we should get more correlations here and a high positive correlation between time of going to bed and time of arising. Another way of checking the present finding would be to compare the variability of rising time with the variability of retiring time. We predict that the latter would be greater than the former, although we have not calculated the former. We took the length of night from our machine records, which give a more reliable value than could be obtained from the parents' reports. Our finding agrees with that of the sleep study of the Institute of Child Welfare (2, page 11), which finds the shorter night sleep in summer due to later bedtime uncompensated for by later rising time. As an application of this finding we may suggest that if one wishes to keep a child in bed longer, it would be a good idea to put him to bed earlier, rather than to rely entirely upon keeping him there later in the morning.

EFFECT OF WEIGHT AND BODY BUILD ON SLEEP

We thought we noticed that some of the rather fat children were sleeping more quietly and longer than the slender ones. We therefore calculated the ratios of each child's weight to his height for each month during the experimental period, each height and weight

TABLE 34.—MEAN AGE AND WEIGHT OF SUBJECTS FOR THE EXPERIMENTAL PERIOD AND MEAN WEIGHT-HEIGHT RATIO

Subject	Mean Age (in months)	Mean Weight (in kilograms)	Mean Weight-Height Ratio	Subject	Mean Age (in months)	Mean Weight (in kilograms)	Mean Weight-Height Ratio
1.	40.3	14.36	.146	12.	47.3	14.42	.148
2.	34.0	15.69	.163	22.	43.5	16.56	.164
3.	32.1	14.69	.156	13.	48.5	15.42	.153
4.	32.0	15.40	.161	14.	35.0	13.71	.153
5.	50.0	17.32	.172	15.	38.6	16.48	.168
6.	48.0	16.03	.161	16.	39.0	15.26	.151
7.	41.5	14.97	.151	18.	44.5	15.70	.163
8.	36.6	15.90	.163	19.	39.5	16.38	.167
9.	46.5	19.06	.187	20.	53.6	18.00	.171
10.	34.1	16.70	.168	21.	53.5	19.19	.183
11.	51.3	14.20	.143	32.	41.3	14.73	.157

being the middle values of three determinations taken by the nursery school physician and nurse. The average of all the ratios for each child is shown in Table 34 along with the average age of the child during the experimental period in which all the data were available. The age range was so small that the weight-height ratios were not correlated for the effect of age. The rank-order correlation between weight and the weight-height ratio is .93, but between either weight or weight-height ratio and either the length or quietness of sleep the correlations are all negligible. The correlation between weight and MRP2 is —.06; between weight and length of sleep, .04; between body build and mean rest period, —.06; and between body build and length of sleep, —.01. The correlation between age and weight is .38, and between age and the weight-height ratio, .14. No relationship was discovered between either weight or body build and sleep.

Effect of Posture on Sleep

We attempted to study the relationship between sleeping posture and the quietness of sleep by having the parents record each evening the position in which the child fell asleep. We were able, however, to obtain a total of only 192 usable records on nine children. Our reason for the small number was that records were usable only on children for whom at least three out of the four positions—back (B), abdomen (A), right side (R), and left side (L)—were reported. For the children who adopted all four postures, the postures are given here in the order of the quietness of sleep on the nights when those postures were recorded. For Subject 3 this order was A, L, R, B; for Subject 4, B, R, L, A; for Subject 10, B, A, R, L; for Subject 12, A, R, B, L; and for Subject 32, R, B, A, L. For those subjects whose records indicate only three positions, the order is as follows: for Subject 21, L, A, B; for Subject 19, B, R, L; for Subject 16, B, L, R; and for Subject 15, B, R, L. All we can suggest is that sleeping on the back and abdomen seems to tend to induce quieter sleep than sleeping on the sides. This finding disagrees with the differences found by Boynton and Goodenough (3, page 275). They found, however, that the length of time required to go to sleep was shorter when the subjects lay on the back and abdomen and longer when they lay on the sides. This problem should be studied more fully as it applies to night sleep, and the findings compared with those on naps, such as are reported by Boynton and Goodenough.

Effect of Ultra-Violet Rays on Sleep

Mrs. Ruby Glockler of the Institute of Child Welfare of the University of Minnesota gave experimental ultra-violet ray treatments to a number of children, including some of our subjects. The mean rest periods that we found for her experimental and control groups are shown in Table 35, along with the number of light treatments and of sleep records. It is interesting to note that while for all the children in the control group there was a decrease in mean rest period between the pre-experimental and the experimental periods (a seasonal variation), for three of the six children who took the treatments the mean rest period was longer during the time the treatments were given, and for those on whom we have data there was a decrease in the mean rest period after the light treatments were discontinued. It is also significant that of the three children (Subjects 6, 10, and 7) whose sleep did not increase in quietness during the treatments, two had only small amounts of treatment, as shown by the total minutes of irradiation. It should be noted further that one of these two boys (Subject 7) was very sensitive to the light and was "sunburned" a number of times, which according

TABLE 35.—LENGTH OF MRP2 FOR CHILDREN WHO RECEIVED AND FOR THOSE WHO DID NOT RECEIVE IRRADIATION TREATMENT

Subject	Pre-Experimental Period		Experimental Period					Post-Experimental Period	
	No. of Nights	MRP2	No. of Treatments	Total Minutes of Treatment	No. of Nights	MRP2		No. of Nights	MRP2
			EXPERIMENTAL GROUP						
6. . .	43	8.77m	8	30	45	7.98m		8	7.57m
10. . .	54	7.52m	13	68	82	7.13m			
4. . .	44	8.39m	12	65	56	8.43m		4	7.66m
3. . .	38	8.53m	13	86	48	9.13m		19	7.26m
7. . .	36	9.09m	14	34	67	8.88m			
11. . .	25	8.92m	5	20	16	9.38m			
			CONTROL GROUP						
8. . .	59	6.85m			31	6.64m			
9. . .	36	8.91m			24	8.90m			
12. . .	49	8.41m			62	7.88m			
22. . .	52	9.72m			50	8.76m			
5. . .	42	8.47m			44	7.96m			

to his mother's reports caused him to be very restless at night. The other boy (Subject 6) was ill for a month during the experimental period. We have no information that might explain the third exception. The number of children is small, but the data very strongly suggest that the ultra-violet ray treatments, given in the winter when there was very little sunshine, tended to produce more quiet sleep or to offset the increase in nocturnal activity that normally comes in December and January. It also suggests that the midwinter restlessness during sleep which we find in our group of children as a whole may be a function of the lack of sufficient ultra-violet rays in the sunshine that filters through the winter smoke clouds of southeast Minneapolis, where all our subjects lived.

SUMMARY

1. No consistent effect of nursery school attendance upon the length or variability of the mean rest period was discovered, although some children slept significantly more quietly on nights following days in the nursery school and others slept less quietly. The total time of sleep tended to be shorter but more regular when the child went to nursery school.

2. Children tended to sleep more quietly on nights when they wet the bed, although there were individual differences.

3. If the child engaged in violent exercise between supper and bedtime, he seemed to sleep a little more quietly.

4. The data suggest that constipation may produce longer sleep and looseness of the bowels more restless sleep.

5. The negative correlation found between time of going to bed and length of stay in bed means that the subjects rose about the same time as usual even though they went to bed late. If they stayed in bed as long on nights when they went to bed late as on nights when they went to bed early, the correlation would have been zero. It is probably easier to regulate the length of sleep by changing the bedtime than by changing the rising time.

6. No relationship was found between either weight or body build and the length and quietness of sleep characteristic of each child.

7. The data suggest that children tend to sleep more quietly when they fall asleep on the back or abdomen than when they fall asleep on either side. This disagrees with previous findings on naps, and the experiment should be repeated with larger numbers of records.

8. Ultra-violet ray treatments tended to increase the quietness of sleep or to offset seasonal decreases. It is suggested that the midwinter decrease in the quietness of sleep is a function of the poverty of ultra-violet rays in the winter sunshine.

VIII. SUMMARY AND CONCLUSIONS

SUMMARY

This study was carried out in a attempt to ascertain through objective means the general character of the sleep of young children and the relationship of various factors to the length and quietness of their sleep. It was hoped thus to establish norms of the distribution of children's nightly activity and of the mean length of time between their successive major changes in position.

For this purpose we used twenty-two children aged from twenty-five to fifty-eight months and attending the nursery school at the Institute of Child Welfare, University of Minnesota. Thirteen of them were studied intensively for five months or more. Each child was supplied with a crib especially designed for the experiment, the beds having attached to the under side of the spring a Johnson kinetograph which recorded the child's motions during the night. The children slept in their own homes. Each child's mother turned on the recording instrument when the child went to bed and turned it off when he got up in the morning. The mothers also kept records of the children's time of going to bed, getting up, etc., on forms provided for this purpose. Usable records were obtained for a total of 3,339 nights.

The night was divided, for purposes of this study, into the initial active period, the rest period, and the final active period, each of these being subdivided into five-minute intervals. The initial active period consisted of the number of five-minute intervals after the child went to bed during each of which at least one movement was recorded by the kinetograph. The rest period was the sum of five-minute intervals starting after the first rest period greater than five minutes and stopping at the beginning of the final active period, which consisted of the period after the child awoke but before he left the bed in the morning, indicated on the kinetograph by rest periods none of which were greater than 2.5 minutes. Each child's mean rest period consisted of the average number of minutes from the middle of one active interval to the beginning of the next.

FINDINGS

1. A much greater similarity was found between the curves showing the nocturnal activity of our child subjects than Johnson found

for his adult subjects. Such individual differences as were found among the children seem to have been due, to some extent at least, to the varied sleeping conditions in the different homes, which ranged from very favorable to very poor.

2. Our group curves of activity during sleeping time tended to confirm the findings of other investigators in indicating that activity during sleep is periodic or rhythmic.

3. A small sex difference in the activity curves of our children was evident, this difference being somewhat greater than that found between those children who had high correlations with the whole group and those who had low correlations. The boys went to sleep more promptly than the girls and slept more quietly throughout the night. The large individual differences, however, prevent this difference from being very significant. The girls were in bed an average of eight minutes longer than the boys, but slept only five minutes longer.

4. The mean rest period for the entire night for all the children was about 7.25 minutes, the normal range being from 5.9 to 8.4 minutes. The mean rest period during the time devoted to sleep was about 8 minutes, the normal range being 6.4 to 9.4 minutes. Mean rest periods for adults, ascertained through previous studies, range from 6 to 22 minutes, with averages from 7 to 13 minutes. Thus the children were shown to move about twice as often as adults. They also took about twice as long as adults to go to sleep, their average initial active period being between 30 and 35 minutes, compared with 10 to 20 minutes for adults.

5. Quietness of the children's sleep for the group as a whole showed no relationship with length of sleep nor with the time of going to bed; i. e., the children did not sleep more quietly when they stayed in bed longer or went to bed earlier. There was some evidence suggesting that night sleep was less quiet if the child had had an afternoon nap. Sleep became more restless toward morning.

6. Excepting the tendency of extreme temperatures to induce more restless sleep, neither room temperature nor outside temperature appeared to have any noticeable effect on nocturnal activity. The children showed a tendency to sleep longer in cold than in warm weather and to sleep more quietly in lower relative humidities. When fever temperatures (100° to 103° F.) were present in any of the children, they tended to sleep a little less quietly than usual, but the difference was too slight to be regarded as significant.

7. Children at three years of age showed more activity during sleep than did children at two or at four years of age. Children aged three years also took longer to go to sleep than did those at the other ages.

8. The children's sleep in February and March was a little more quiet than the average for the year, and in April, May, and June it was a little more restless than the average.

9. The average time in bed for all the children was about eleven hours.

10. Quietness of sleep at night showed no consistent effect of the children's attendance or non-attendance at nursery school during the preceding day.

11. Children tended to sleep more quietly on nights when the bed was reported wet, when they had engaged in violent exercise between supper and bedtime, when they fell asleep lying on the back or abdomen rather than on the side, and when they had had ultra-violet ray treatments during the day. The data suggest that constipation may produce longer sleep and looseness of the bowels more restless sleep. No relationship was found between either weight or body build and the length or quietness of sleep.

Suggestions for Future Study

1. An observational study designed to discover whether the first rest period greater than five minutes during the night actually co-incides with the beginning of what we ordinarily call "sleep." (The cessation of activity for at least five consecutive minutes was taken as the criterion of sleep in this study.)

2. A study using a larger number of children than were included in the present investigation, and designed to discover whether the sex differences that we found in the nocturnal motility curves are actually significant.

3. A further study of the relationship between length of sleep and quietness of sleep at various age levels; also the relationship between day naps and night sleep.

4. A further study of the relationship between sleep and (a) room temperature and (b) relative humidity, the latter factors being much more rigidly controlled than was possible in this study, and a larger number of subjects being used.

5. A further study of the effect of body temperatures on sleep.

6. A further study designed to discover whether the greater

restlessness of three-year-old children, as found in this and other studies, is a true difference.

7. A study of the relationship between the diurnal and the nocturnal activity of children of various ages.

8. Further study, using a larger number of subjects, of the effects on sleep of the various factors referred to in Chapter VII of this study.

APPENDIX

SUMMARY OF NON-STATISTICAL DATA FOR EACH SUBJECT

Subject 1.—Girl; slender build; quiet, timid, and shy; sometimes called a nervous child. Father a professor of engineering. Date of child's birth, March 2, 1925. Experimental period, March 22, to December 22, 1928.

This child lived, at the time of the experiment, in a fine new house on a quiet street. The sleeping room she occupied was in a part of the house away from the street, and was large and well ventilated and heated. The bed was always just a few inches from the wall, and other details were attended to with equal care. The child was put to bed at almost exactly the same time each evening, the bed arranged with great care, and the child left immediately. The mother was a perfect housekeeper, very careful and precise in her work, especially that which concerned the care and welfare of the child. Both parents were very quiet at all times, spoke in low tones, and never encouraged the child to be boisterous or anything but well behaved. Her life was ordered and supervised with great regularity and attentiveness.

Subject 2.—Girl; had speech impediment. Parents well educated. Father a traveling representative of a publishing house. One older boy. Date of child's birth, June 18, 1925. Experimental period, March 7 to August 13, 1928.

This child lived on the second floor of a duplex near the intersection of two rather quiet streets. The sleeping room, which she shared with her older brother, was segregated from the rest of the house and on the side away from the street. The room was small, and the two beds and dresser which it contained made it rather crowded. Ventilation, from one window very close to the foot of the bed and another near the head, was fair. The bed was well taken care of. The house was not noisy.

Subject 3.—Girl. Father a professor in the medical school. Mother a college graduate. Date of child's birth, January 12, 1926. Experimental period, February 24, 1928, to April 26, 1929.

This child's family moved during the experimental period. First they lived in a large house near the intersection of two rather busy streets, but the traffic was not heavy after dark. The child had a well-ventilated private room on the second floor away from the street. The second house was the lower apartment in a duplex on a quiet residential street, one block from the interurban bus and streetcar line. The sleeping room was on the ground floor near a private driveway. It was shared by the child's mother, but since the room was very large, this fact probably occasioned very little disturbance. There were three windows, at least one of which was nearly always open, since the house was slightly overheated. The bed was well cared for.

Subject 4.—Boy; active. Father chief engineer of machinery in a large feed mill. Three older girls. Date of child's birth, January 7, 1926. Experimental period, February 23, 1928, to April 20, 1929.

This child's house was situated only half a block from an interurban highway, but on a fairly quiet street. The sleeping room, which was large and well ventilated, was shared by the child's parents. The house was not noisy, for all the children were exceptionally quiet and well behaved.

Subject 5.—Boy. Father owner of a small wholesale and retail grocery. One older girl. Date of child's birth, August 30, 1924. Experimental period, March 16, 1928, to April 3, 1929.

The child's room was on the first floor of a new house on a quiet street. The room, away from the street and rather isolated from the rest of the house, was shared by the older child, who, however, was very quiet. Heat and ventilation were good.

Subject 6.—Boy; very lively. Father an official in large wholesale creamery; had suffered partial shell shock during the World War. Mother not well, and slightly annoyed by the children's boisterous play. Younger boy and baby twin girls in family. Date of child's birth, August 14, 1924. Experimental period, January 30, 1928, to April 11, 1929.

The house, which was new, was about a block from the heavy freight line of a railroad. The sleeping room, on the second floor overlooking a quiet street, was also occupied by a maid who took care of the children, and with whom they were frequently left alone while the parents went out. The child was not well trained and often got up again after being put to bed very early.

Subject 7.—Boy; very precocious. Father a college professor. Date of child's birth, March 18, 1925. Experimental period, January 31, 1928, to April 12, 1929.

The house where this child lived was on a rather noisy street, but his sleeping room was on the side away from the street, overlooking a large lawn. The room, on the first floor, was occupied by the child alone. It was well heated and ventilated. There were no other children in the house. A quiet elderly couple lived on the floor above. The child was well cared for but not coddled. He was left alone a good deal by his parents, who were personally well adjusted. His bed was well cared for.

Subject 8.—Boy. Father an official of an oil company. One older boy in the family. Date of child's birth, July 21, 1925. Experimental period, February 1, 1928, to February 15, 1929.

The house was on a rather noisy street intersection, and the child's room, on the second floor, was close to this street. A thin partition separated this room from a corridor and the bathroom. No one else slept in the room, which was very small and had only one window. A certain amount of loud speaking was occasioned by the fact that the grandfather, who lived with the family, was slightly deaf.

Subject 9.—Boy; fat. Father a milkman. Two older girls in family. Date of child's birth, October 29, 1924. Experimental period, February 8, 1928, to March 11, 1929.

The family occupied the first floor of a small house on a fairly quiet street. The child's sleeping room was in the back of the house away from the street but about a block from a freight transfer line. The room, which was shared by the child's parents, was very small, and though often cold, was never well ventilated in cold weather. The older children usually slept in the next room but were often put in this room when there was a party at the house. Parties were not infrequent, sometimes large for the size of the house, and often continued till a late hour, despite the fact that the father rose at 4 A.M. to go to work.

The child was put to bed at various hours to suit the convenience of the parents, being sent to bed earlier than usual when they wished to go out in the evening and later than usual on other occasions. He was also taken up at irregular times in the morning. If the parents were out late at night, the child might be left in bed very late the next morning. This often happened on Saturdays and Sundays, when there was no nursery school. Supper time was fairly regular, but the diet was not well controlled, and any meal might be made up entirely of heavy, rich foods.

This child's bed was often wet, even until afternoon. This meant not only that the child was probably not dried during the night when the accident occurred, but also that the bed was not thoroughly dried even the next day. Although the mother's daily reports on enuresis indicated that the bed was wet only fourteen nights during thirteen months, the bed was frequently observed by the experimenter to be wet in the daytime, and the springs were rusted.

Subject 10.—Girl; bashful. Father a milkman. Older boy and girl in family. Date of child's birth, November 10, 1925. Experimental period, February 8, 1928, to April 15, 1929.

This child's house was on the same street as that of Subject 9. It was the same distance from the railroad but less than a block from the interurban bus and streetcar line, and only about a hundred yards from the intersection of the railroad and streetcar lines. The child's sleeping room was on the side of the house away from these traffic arteries, however, and was also shielded from the more quiet street on which the house stood. The house itself, which was quiet, stood much farther from the street than the nearby houses. The sleeping room, on the second floor, was large and well heated and ventilated. The mother, who slept across the corridor, was very quiet. The child's bed was well cared for.

Subject 11.—Girl; very slender and small for her age. Father owner of a tire shop. Older girl and baby girl in family. Date of child's birth, June 6, 1924. Experimental period, February 11, 1928, to April 4, 1929.

The family lived on the second floor of a house on a quiet street. The room was not large for the three beds it contained. Another child and a maid slept in the same room. The child was put to bed at any time that suited the convenience of the parents, who often went out or had friends in to play bridge. The child was likewise taken up at varying times in the morning. She frequently got into her parents' bed in the night or early morning and was allowed to remain there; or her own sleep was disturbed by her older sister's getting into the same bed with her. The bed was often not well aired and made up.

Subject 12.—Boy; very active. Father treasurer of a large lumber plant. Older boy and two older girls in home. Date of child's birth, September 12, 1924. Experimental period, February 12, 1928, to April 2, 1929.

The child's room, which was large and well ventilated, was on the second floor of a new house, overlooking a quiet street. The parents slept in the same room. The mother had insufficient time to care for all her children in addition to doing all her own housework and baking. The child was not well trained temperamentally nor as regards sleeping habits. The records show that the bed was wet 151 nights and dry 71.

Subject 22.—Girl; heavy for her age. Father an apartment engineer and repairman. Date of child's birth, May 26, 1925. Experimental period, October 16, 1928, to March 15, 1929.

This child slept in a large, well-heated, and well-ventilated private room on the second floor of a new house on a quiet street about half a block from a streetcar line having infrequent service. The house was quiet and the habits of the family regular. The child was well cared for, the bed always well made. The enuresis report shows that the bed was wet 19 times and dry 29 times in 116 days, 68 days being left blank.

Subject 13.—Girl; very slender. Father a college professor. One older girl in home. Date of child's birth, November 5, 1925. Experimental period, November 16 to December 15, 1929.

Subject 14.—Boy. Father a railway airbrake inspector and repairman. Date of child's birth, January 13, 1927. Experimental period, November 13, 1929, to January 26, 1930.

Subject 15.—Boy. Father a fire insurance agent. Date of child's birth, May 3, 1926. Experimental period, April 25 to December 22, 1929.

Subject 16.—Boy. Father works in a laboratory. Mother was a teacher before marriage. One older girl in family. Date of child's birth, February 12, 1926. Experimental period, April 29 to June 10, 1929.

Subject 18.—Boy; very quiet. Father owner of a barber shop. One older boy and one older and one younger girl in family. Date of child's birth, December 17, 1925. Experimental period, April 26, 1929, to January 21, 1930.

Subject 19.—Boy. Father a manufacturer. Mother a high school teacher before marriage. One older girl in family. Date of child's birth, January 17, 1926. Experimental period, April 5 to May 23, 1929.

Subject 20.—Boy; backward. Father a machine draftsman. Date of child's birth, February 1, 1925. Experimental period, April 3 to December 29, 1929.

Subject 21.—Boy. Father a railway brakeman. One older boy in family. Date of child's birth, March 18, 1925. Experimental period, May 1 to December 22, 1929.

Subject 32.—Girl; very slender. Father contracting manager of a foundry company. Two older girls in family. Date of child's birth, March 14, 1926. Experimental period, April 22 to December 26, 1929.

Summary of Statistical Data for Each Subject

Date	No. of Nights	Time in Bed Mean	Time in Bed σ	MRP1 Mean	MRP1 σ	Time Asleep Mean	Time Asleep σ	MRP2 Mean	MRP2 σ	Initial Active Period No. of Nights	Initial Active Period Mean	Initial Active Period σ
SUBJECT 1												
Mar, 1928.	10	11h 24m	19.0	5.07m	0.24	10h 33m	36.0	5.39m	0.66	10	41.5m	24.0
April, 1928. . . .	30	11h 34m	27.0	6.19m	0.84	10h 52m	41.5	6.65m	0.94	30	30.0m	21.5
May, 1928.	31	11h 42m	24.0	6.18m	0.63	10h 56m	43.0	6.68m	0.71	31	37.9m	21.1
June, 1928. . . .	29	11h 09m	32.0	5.73m	0.57	10h 19m	38.0	6.25m	0.57	29	47.4m	25.7
July, 1928. . . .	30	10h 53m	19.5	5.67m	0.63	9h 44m	42.5	6.37m	0.82	30	59.5m	16.5
Aug, 1928. . . .	26	10h 41m	37.0	5.59m	0.73	9h 44m	42.5	6.14m	0.96	27	41.3m	13.9
Sept, 1928. . .	19	10h 46m	28.0	5.69m	0.66	9h 56m	46.0	6.12m	0.62	19	37.9m	18.5
Oct, 1928. . . .	25	11h 12m	24.0	6.22m	0.79	10h 35m	35.5	6.62m	0.94	26	31.9m	16.0
Nov, 1928. . . .	29	11h 32m	24.0	6.04m	0.35	10h 33m	40.5	6.52m	0.79	29	38.3m	17.9
Dec., 1928. . .	21	11h 31m	29.5	6.08m	0.59	10h 33m	35.0	6.56m	0.57	21	43.8m	17.6
All months. . .	250	11h 13m	33.5	5.90m	0.74	10h 23m	44.5	6.40m	0.85	252	41.1m	21.3
SUBJECT 2												
Mar, 1928. . . .	22	11h 30m	37.5	8.27m	1.29	10h 32m	102.5	9.58m	1.83	23	50.7m	27.1
April, 1928. . .	28	11h 41m	45.0	7.58m	1.05	10h 30m	49.0	8.94m	1.36	28	55.7m	27.2
May, 1928. . . .	30	11h 27m	57.5	7.56m	1.10	10h 21m	53.5	8.75m	1.32	30	52.5m	33.3
June, 1928. . .	17	11h 45m	32.5	7.98m	1.22	10h 56m	77.0	8.90m	1.23	18	41.4m	37.0
Aug, 1928. . .	4	10h 55m	15.0	10.31m	0.87	10h 29m	9.0	11.26m	0.87	4	23.8m	10.9
All months. . .	101	11h 33m	42.0	7.90m	1.28	10h 32m	69.0	9.11m	1.52	103	49.9m	31.0
SUBJECT 3												
Feb., 1928. . .	3	11h 15m	49.5	10.54m	1.19	10h 34m	25.0	11.62m	1.05	3	26.7m	18.9
Mar., 1928. . .	20	10h 45m	35.0	8.18m	1.66	10h 21m	41.5	8.79m	1.88	23	20.9m	17.0
April, 1928. . .	27	10h 39m	32.0	7.92m	1.98	10h 16m	33.5	8.42m	2.20	27	18.2m	10.9
May, 1928. . . .	29	10h 51m	36.0	8.09m	1.23	10h 19m	29.0	8.83m	1.49	30	25.2m	12.5

Summary of Statistical Data for Each Subject—*Continued*

Date	No. of Nights	Time in Bed Mean	σ	MRP1 Mean	σ	Time Asleep Mean	σ	MRP2 Mean	σ	Initial Active Period No. of Nights	Mean	σ
June, 1928.	28	10h 11m	62.0	7.96m	1.55	9h 34m	51.0	8.79m	1.71	28	32.7m	16.3
July, 1928.	27	9h 55m	59.0	7.93m	1.32	9h 23m	38.0	8.66m	1.31	29	21.7m	16.5
Aug., 1928.	25	10h 05m	51.5	7.84m	1.46	9h 29m	38.5	8.61m	1.71	25	21.0m	13.7
Oct., 1928.	17	11h 08m	62.5	7.21m	1.17	9h 58m	59.5	8.39m	1.40	17	18.6m	12.1
Nov., 1928.	22	11h 42m	39.5	7.80m	1.32	10h 47m	47.5	8.83m	1.48	24	27.9m	14.4
Dec., 1928.	15	11h 22m	41.0	7.16m	1.12	10h 30m	43.0	7.99m	1.31	15	31.7m	10.0
Jan., 1929.	10	11h 16m	36.0	8.94m	1.08	10h 38m	39.5	10.32m	1.33	11	17.3m	12.2
Feb., 1929.	17	11h 18m	34.5	8.43m	0.75	10h 42m	41.0	9.24m	0.80	17	19.4m	7.3
Mar., 1929.	17	11h 07m	31.5	7.78m	0.91	10h 35m	27.0	8.40m	0.98	17	18.8m	13.6
April, 1929.	23	11h 06m	32.0	7.02m	0.77	10h 33m	28.5	7.53m	0.92	23	27.9m	14.9
All months.	280	10h 48m	50.0	7.87m	1.46	10h 10m	51.0	8.65m	1.63	289	23.6m	14.7
SUBJECT 4												
Feb., 1928.	6	11h 47m	28.5	9.15m	1.89	11h 04m	24.0	10.31m	2.37	6	36.7m	18.7
Mar., 1928.	24	11h 15m	34.0	7.99m	0.93	10h 47m	35.5	8.6om	1.21	26	23.1m	18.7
April, 1928.	24	11h 09m	46.0	7.61m	0.96	10h 34m	54.0	8.31m	1.18	24	27.9m	18.9
May, 1928.	22	10h 54m	77.5	7.39m	1.20	10h 25m	71.0	7.93m	1.49	24	26.1m	18.5
June, 1928.	24	10h 54m	91.5	7.09m	1.16	10h 10m	78.0	7.69m	1.32	25	40.6m	49.1
July, 1928.	22	10h 10m	73.0	8.72m	1.51	9h 55m	24.0	9.19m	1.54	21	16.0m	43.8
Aug., 1928.	19	10h 46m	5.0	8.07m	0.97	10h 18m	48.0	8.70m	1.23	18	21.4m	42.4
Sept., 1928.	8	9h 37m	62.0	9.11m	1.68	9h 20m	58.0	9.68m	1.91	7	13.6m	9.2
Oct., 1928.	15	9h 49m	67.5	7.88m	1.71	9h 18m	60.5	8.51m	1.69	15	28.0m	18.1
Nov., 1928.	17	11h 21m	32.0	7.65m	0.96	10h 40m	51.5	8.45m	1.21	17	33.0m	31.2
Dec., 1928.	19	10h 52m	65.5	7.49m	0.98	10h 15m	60.0	8.14m	1.00	19	29.2m	18.2
Jan., 1929.	17	11h 36m	79.0	8.04m	1.22	10h 54m	89.0	8.92m	1.58	16	41.3m	26.4
Feb., 1929.	16	11h 16m	55.0	7.86m	0.92	10h 46m	49.5	8.44m	0.91	17	27.1m	18.7

Month												
Mar., 1929	21	10h 43m	65.5	7.38m	1.17	10h 01m	53.0	8.12m	1.22	21	27.9m	29.8
April, 1929	12	10h 32m	62.0	7.59m	1.15	9h 53m	42.0	8.34m	1.54	12	36.7m	34.8
All months	266	10h 52m	71.5	7.99m	0.58	10h 19m	62.0	8.48m	1.46	268	28.5m	26.5

SUBJECT 5

Month												
Mar., 1928	6	10h 45m	38.0	7.64m	1.30	10h 11m	61.0	8.48m	1.79	7	22.9m	18.9
April, 1928	26	10h 42m	40.0	7.31m	1.24	9h 56m	61.5	8.10m	1.44	28	33.4m	22.1
May, 1928	19	10h 16m	33.5	7.18m	0.96	9h 31m	39.0	7.89m	1.52	22	32.5m	20.9
June, 1928	30	10h 10m	46.0	7.68m	1.35	9h 37m	53.5	8.34m	0.36	30	27.2m	20.3
July, 1928	23	10h 13m	42.5	7.56m	1.17	9h 43m	48.5	8.21m	1.11	24	25.7m	20.1
Aug., 1928	11	10h 21m	49.0	7.70m	0.91	10h 05m	53.0	8.12m	1.00	11	15.5m	12.4
Oct., 1928	19	10h 30m	56.0	7.48m	1.03	9h 29m	68.5	8.15m	1.02	21	22.9m	23.9
Nov., 1928	18	9h 57m	50.5	7.80m	0.88	9h 28m	36.5	8.50m	0.85	18	21.7m	17.8
Dec., 1928	17	9h 40m	56.0	7.99m	0.96	9h 24m	47.5	8.46m	0.96	17	13.6m	20.9
Jan., 1929	17	9h 30m	46.0	8.06m	0.99	9h 13m	41.5	8.47m	1.05	17	7.4m	7.3
Feb., 1929	16	9h 29m	57.5	7.59m	1.29	9h 14m	51.5	7.94m	1.48	16	8.8m	15.1
Mar., 1929	16	8h 51m	44.0	7.03m	0.64	8h 21m	51.0	7.50m	0.60	20	4.8m	5.6
April, 1929	2	10h 10m	5.0	8.86m	0.86	9h 55m	0.3	9.26m	0.74	2	2.5m	0.3
All months	220	10h 00m	55.5	7.58m	1.16	9h 30m	55.0	8.18m	1.19	233	20.9m	20.8

SUBJECT 6

Month												
Jan., 1928	2	12h 18m	2.5	7.01m	0.64	11h 30m	32.5	7.61m	0.25	2	45.0m	35.0
Feb., 1928	16	11h 35m	31.5	7.05m	0.88	10h 56m	46.5	7.73m	0.76	17	24.7m	31.0
Mar., 1928	19	12h 19m	50.5	7.41m	1.14	11h 32m	62.5	8.15m	1.16	20	29.3m	28.1
April, 1928	20	11h 52m	185.5	7.51m	0.83	10h 45m	31.5	8.67m	0.41	22	24.8m	54.2
May, 1928	30	11h 34m	35.5	7.13m	0.71	10h 20m	65.0	8.30m	0.71	30	62.4m	32.7
June, 1928	22	12h 15m	33.5	7.20m	0.84	11h 10m	41.5	7.99m	0.93	23	39.8m	29.0
July, 1928	19	11h 24m	58.5	7.52m	0.86	10h 39m	66.0	8.33m	0.75	20	36.0m	27.3
Aug., 1928	16	11h 23m	57.0	7.41m	0.95	10h 12m	72.5	8.73m	1.01	16	58.5m	33.5
Sept., 1928	8	11h 08m	64.0	8.10m	0.55	10h 31m	69.5	8.95m	0.85	8	25.7m	18.1
Oct., 1928	19	11h 57m	52.0	7.33m	0.81	10h 42m	51.0	8.46m	0.90	21	51.1m	30.8
Nov., 1928	17	11h 58m	53.5	7.54m	0.44	11h 09m	52.0	8.30m	0.46	19	34.2m	31.1
Dec., 1928	19	11h 57m	53.5	7.77m	1.13	10h 49m	77.0	9.56m	1.79	19	64.8m	62.7

Summary of Statistical Data for Each Subject—*Continued*

Date	No. of Nights	Time in Bed Mean	σ	MRP1 Mean	σ	Time Asleep Mean	σ	MRP2 Mean	σ	Initial Active Period No. of Nights	Mean	σ
Jan., 1929. . . .	19	12h 44m	84.5	7.20m	0.74	12h 35m	34.5	7.76m	0.60	22	38.7m	23.7
Feb., 1929. . . .	12	12h 44m	32.5	6.99m	0.69	11h 54m	19.5	7.68m	0.77	12	22.9m	11.9
Mar., 1929. . . .	23	12h 20m	58.0	7.07m	0.69	11h 06m	19.5	8.24m	1.12	23	47.6m	46.3
April, 1929. . . .	6	12h 12m	37.0	6.68m	0.69	11h 00m	34.5	7.53m	0.44	6	46.7m	47.4
All months. . .	267	11h 58m	73.0	7.32m	0.88	10h 58m	62.5	8.31m	1.11	280	43.1m	36.8
SUBJECT 7												
Jan., 1928. . . .	1	12h 05m	...	7.50m	...	12h 03m	...	7.57m	...	1	0.0m	...
Feb., 1928. . . .	21	11h 11m	64.5	7.37m	1.02	10h 44m	63.5	7.54m	2.54	25	20.4m	16.0
Mar., 1928. . . .	25	11h 10m	46.5	7.08m	0.88	10h 42m	48.0	7.54m	0.95	26	22.3m	15.0
April, 1928. . . .	25	10h 34m	62.5	7.23m	1.10	9h 56m	63.5	7.87m	1.21	26	33.5m	23.2
May, 1928. . . .	26	11h 01m	57.5	7.69m	1.28	10h 17m	64.0	8.54m	1.20	27	42.4m	33.7
June, 1928. . . .	19	11h 02m	68.5	7.48m	0.77	10h 17m	68.0	8.16m	0.79	21	26.2m	23.0
July, 1928. . . .	13	10h 49m	56.0	6.83m	0.64	10h 07m	58.5	7.59m	0.80	14	31.1m	16.1
Aug., 1928. . . .	5	11h 21m	42.5	6.96m	1.05	10h 19m	76.5	7.65m	0.71	5	47.0m	38.2
Oct., 1928. . . .	21	11h 30m	37.0	8.23m	1.24	10h 42m	43.5	8.72m	1.18	26	19.4m	18.7
Nov., 1928. . . .	14	11h 11m	34.0	7.99m	0.60	10h 32m	34.5	8.81m	0.72	21	34.5m	17.4
Dec., 1928. . . .	14	11h 01m	35.5	8.49m	3.06	10h 11m	50.5	9.63m	0.93	21	39.8m	26.4
Jan., 1929. . . .	18	11h 25m	86.5	8.31m	1.26	10h 33m	89.5	8.94m	1.17	21	19.3m	28.7
Feb., 1929. . . .	22	10h 59m	43.0	8.37m	0.87	10h 17m	70.5	9.28m	0.78	24	34.2m	40.6
Mar., 1929. . . .	26	10h 39m	62.0	7.89m	0.88	10h 12m	67.0	8.45m	0.85	26	24.1m	18.1
April, 1929. . . .	9	10h 23m	52.5	8.24m	1.64	9h 54m	44.5	8.82m	1.55	10	22.0m	17.6
All months. . .	259	10h 57m	60.5	7.73m	1.24	10h 23m	61.0	8.41m	1.18	294	28.8m	25.9
SUBJECT 8												
Feb., 1928. . . .	25	11h 20m	31.0	7.07m	0.73	10h 53m	23.5	7.50m	0.85	26	22.7m	9.6

Mar., 1928	25	11h 35m	36.5	7.03m	0.74	11h 04m	32.5	7.47m	0.75	26	26.4m	11.8
April, 1928	29	11h 08m	40.0	6.39m	0.85	10h 25m	66.0	6.90m	1.17	29	29.2m	14.6
May, 1928	25	10h 57m	63.0	6.61m	1.02	10h 22m	40.0	7.13m	1.10	25	30.8m	14.7
June, 1928	11	11h 02m	71.0	6.14m	0.56	10h 20m	75.5	6.61m	0.72	11	38.7m	17.8
Oct., 1928	23	11h 03m	34.0	6.13m	0.48	10h 22m	31.0	6.61m	0.54	23	38.7m	7.5
Nov., 1928	21	11h 13m	40.5	6.40m	0.78	10h 30m	36.0	6.97m	0.88	21	42.6m	15.8
Dec., 1928	29	11h 15m	39.0	6.28m	0.98	10h 23m	43.5	7.03m	0.96	29	45.9m	19.3
Jan., 1929	25	11h 36m	36.0	5.98m	0.43	10h 47m	40.5	6.48m	0.47	25	40.6m	15.3
Feb., 1929	15	11h 21m	49.0	6.02m	0.53	10h 37m	51.0	6.55m	0.58	15	39.0m	10.2
All months	228	11h 15m	48.5	6.44m	0.83	10h 35m	38.0	6.96m	0.96	230	34.9m	16.1

SUBJECT 9

Feb., 1928	18	11h 49m	82.0	7.69m	1.24	10h 26m	68.5	9.17m	1.60	19	57.1m	49.7
Mar., 1928	23	12h 24m	50.0	6.63m	0.96	10h 36m	70.5	8.12m	1.04	25	65.4m	40.9
April, 1928	22	11h 54m	46.0	7.28m	1.02	10h 38m	48.5	8.65m	1.19	25	62.4m	39.5
May, 1928	17	11h 50m	72.5	7.17m	1.38	10h 28m	55.0	8.41m	1.48	17	55.3m	46.7
June, 1928	14	11h 38m	85.0	7.82m	1.08	10h 51m	104.5	8.67m	1.34	18	38.4m	21.3
July, 1928	10	10h 58m	50.0	8.65m	1.04	10h 36m	54.0	9.21m	1.25	10	16.5m	9.3
Aug., 1928	15	12h 58m	53.0	8.01m	0.94	11h 47m	54.0	9.33m	1.32	17	30.6m	23.6
Sept., 1928	4	13h 05m	43.0	7.25m	0.82	10h 57m	23.0	8.95m	0.87	4	50.0m	47.1
Oct., 1928	15	12h 19m	66.5	7.78m	1.02	11h 02m	56.0	9.26m	1.36	15	38.4m	28.4
Nov., 1928	17	12h 49m	81.5	7.60m	0.90	11h 01m	59.0	9.53m	0.90	18	43.9m	32.5
Dec., 1928	15	12h 47m	70.5	7.07m	1.11	11h 02m	55.0	8.55m	1.16	15	61.4m	40.3
Jan., 1929	12	12h 36m	56.5	7.29m	1.28	11h 13m	70.0	8.63m	1.46	12	62.9m	29.1
Feb., 1929	13	12h 50m	73.0	7.04m	0.69	11h 12m	36.0	8.64m	1.21	13	34.6m	13.9
Mar., 1929	5	12h 01m	32.0	7.77m	1.24	10h 46m	14.5	9.06m	0.99	5	44.0m	26.0
All months	200	12h 16m	73.5	7.44m	1.17	10h 52m	65.5	8.82m	1.36	213	48.9m	37.9

SUBJECT 10

Feb., 1928	15	12h 13m	41.0	6.89m	1.02	11h 43m	48.0	7.34m	1.42	17	24.1m	18.9
Mar., 1928	13	11h 30m	64.0	7.41m	1.30	10h 42m	50.5	8.17m	1.49	13	28.1m	17.4
April, 1928	27	12h 10m	11.0	6.72m	0.91	11h 19m	44.5	7.35m	1.00	27	30.4m	22.9
May, 1928	29	11h 15m	57.0	7.00m	0.83	10h 36m	49.0	7.56m	1.00	29	30.9m	19.6

Summary of Statistical Data for Each Subject—*Continued*

Date	No. of Nights	Time in Bed Mean	σ	MRP1 Mean	σ	Time Asleep Mean	σ	MRP2 Mean	σ	Initial Active Period No. of Nights	Mean	σ
June, 1928	10	11h 19m	84.0	6.74m	0.57	10h 13m	68.0	7.73m	1.04	11	55.9m	40.3
July, 1928	19	11h 45m	74.5	6.55m	1.19	10h 55m	39.5	7.15m	1.26	20	48.0m	40.7
Aug., 1928	18	11h 51m	53.5	6.12m	0.39	10h 46m	42.5	6.78m	0.51	18	45.0m	27.0
Sept., 1928	2	12h 20m	20.0	6.77m	0.16	10h 58m	49.0	7.98m	0.69	2	52.5m	27.5
Oct., 1928	22	11h 36m	52.5	7.19m	0.92	10h 53m	53.5	7.84m	1.07	22	26.4m	14.3
Nov., 1928	27	12h 04m	47.5	6.81m	0.81	11h 00m	42.5	7.61m	1.30	27	38.7m	26.8
Dec., 1928	21	12h 16m	68.0	6.70m	0.73	11h 17m	74.5	7.42m	0.87	21	48.1m	33.5
Jan., 1929	21	12h 00m	55.5	6.51m	0.68	10h 58m	60.5	7.25m	0.76	21	51.5m	36.6
Feb., 1929	26	11h 37m	58.0	6.49m	0.67	10h 42m	57.0	7.19m	0.97	27	43.9m	25.1
Mar., 1929	27	12h 17m	19.5	6.37m	0.77	11h 25m	47.5	6.95m	0.74	28	34.7m	27.7
April, 1929	14	11h 56m	39.5	6.54m	0.86	10h 51m	60.5	7.34m	0.90	15	53.7m	66.3
All months	291	11h 52m	62.0	6.71m	0.87	10h 59m	57.0	7.38m	1.10	298	39.1m	32.2
SUBJECT 11												
Feb., 1928	6	9h 48m	77.0	8.77m	1.51	9h 15m	63.0	9.66m	1.75	8	15.0m	10.0
Mar., 1928	21	10h 49m	72.0	7.69m	0.95	9h 58m	69.5	8.72m	1.43	24	25.2m	17.9
April, 1928	17	10h 02m	53.5	8.02m	1.06	9h 25m	76.5	8.87m	1.39	22	20.0m	15.1
May, 1928	12	11h 57m	94.5	7.54m	1.04	9h 12m	60.0	9.35m	0.99	13	37.7m	39.9
June, 1928	9	10h 07m	56.5	7.49m	1.44	9h 22m	57.5	8.23m	1.29	9	46.7m	12.5
July, 1928	3	10h 49m	57.5	7.67m	1.68	9h 21m	45.5	9.27m	1.65	5	50.0m	37.4
Aug., 1928	4	9h 52m	89.0	7.99m	2.37	9h 19m	61.0	8.43m	2.22	5	21.0m	23.6
Oct., 1928	3	9h 40m	57.0	8.37m	0.68	8h 59m	66.5	9.62m	0.33	3	38.4m	38.0
Nov., 1928	2	12h 23m	82.5	6.72m	0.11	9h 30m	47.5	9.93m	0.76	3	60.0m	53.1
Dec., 1928	10	11h 09m	80.5	7.89m	1.75	10h 10m	79.0	8.60m	1.52	10	27.0m	25.0
Jan., 1929	8	10h 38m	57.0	6.99m	0.90	9h 20m	60.5	8.28m	0.90	10	37.0m	32.4
Feb., 1929	5	11h 02m	55.5	8.95m	0.92	10h 33m	32.5	9.72m	1.10	7	15.7m	10.2

Mar., 1929. . . .	15	10h 26m	74.0	8.61m	1.12	9h 55m	80.0	9.42m	1.03	17	21.8m	18.5
April, 1929. . . .	1	10h 30m	...	6.98m	...	9h 58m	...	7.50m	...	1	30.0m	...
All months. . .	116	10h 33m	76.0	7.91m	1.36	9h 39m	69.5	8.95m	1.40	137	28.3m	26.3
					SUBJECT 12							
Feb., 1928. . . .	10	10h 52m	36.0	7.39m	0.95	10h 08m	41.0	8.01m	0.97	13	32.3m	14.4
Mar., 1928. . . .	26	10h 48m	35.5	8.07m	1.15	10h 09m	37.0	8.94m	1.33	26	23.7m	15.1
April, 1928. . . .	26	10h 55m	26.0	7.03m	0.95	10h 01m	35.0	7.88m	1.04	27	33.9m	21.0
May, 1928. . . .	27	10h 45m	32.0	7.26m	1.08	10h 18m	30.5	7.71m	1.08	27	25.2m	22.6
June, 1928. . . .	24	10h 42m	35.5	7.69m	0.86	10h 13m	40.0	8.18m	0.73	24	23.8m	23.1
July, 1928. . . .	20	10h 20m	39.0	8.29m	1.18	10h 07m	18.0	8.60m	1.23	20	10.5m	10.1
Aug., 1928. . . .	19	10h 44m	25.5	8.05m	0.72	10h 27m	36.0	8.38m	0.80	19	6.1m	10.1
Sept., 1928. . . .	8	10h 29m	51.0	8.95m	0.82	10h 10m	63.0	9.47m	0.60	8	4.4m	4.5
Oct., 1928. . . .	13	11h 12m	46.0	7.73m	1.34	9h 43m	38.5	8.24m	1.26	14	19.7m	3.0
Nov., 1928. . . .	10	11h 21m	29.5	7.23m	1.44	10h 22m	28.5	8.33m	0.93	10	40.5m	12.3
Dec., 1928. . . .	26	11h 14m	36.0	7.43m	1.15	10h 21m	46.5	8.32m	1.24	26	36.0m	20.6
Jan., 1929. . . .	28	11h 02m	45.0	8.13m	1.01	9h 47m	63.5	8.34m	0.97	28	51.5m	31.4
Feb., 1929. . . .	22	11h 09m	37.5	7.03m	0.88	10h 12m	37.0	7.90m	0.84	22	39.1m	28.9
Mar., 1929. . . .	24	11h 04m	39.5	7.01m	1.02	10h 19m	32.5	7.72m	1.15	24	27.2m	22.2
April, 1929. . . .	1	10h 20m	...	7.58m	...	9h 48m	...	8.13m	...	1	25.0m	16.8
All months. . .	284	10h 51m	40.5	7.52m	1.10	10h 10m	42.0	8.23m	1.11	289	28.2m	...
					SUBJECT 22							24.0
Oct., 1928. . . .	14	10h 28m	57.0	9.27m	1.07	10h 02m	47.5	10.08m	1.24	14	23.6m	17.4
Nov., 1928. . . .	27	10h 27m	42.5	8.96m	1.05	9h 54m	49.0	9.92m	1.18	27	28.5m	21.5
Dec., 1928. . . .	16	10h 50m	84.5	8.00m	1.10	9h 46m	36.5	9.34m	1.34	20	45.3m	44.8
Jan., 1929. . . .	22	10h 39m	39.5	7.93m	1.00	9h 50m	45.5	8.98m	0.98	22	26.7m	26.7
Feb., 1929. . . .	23	11h 01m	43.5	7.90m	1.13	10h 20m	38.0	8.65m	1.15	23	23.1m	23.1
Mar., 1929. . . .	14	10h 42m	32.0	8.57m	1.32	10h 13m	32.0	9.27m	1.35	14	19.1m	19.1
All months. . .	116	10h 41m	50.0	8.41m	1.24	10h 00m	48.5	9.35m	1.30	120	28.5m	28.5
					SUBJECT 13							
Nov., 1929. . . .	11	11h 47m	28.5	7.38m	0.86	11h 08m	26.5	7.98m	0.80	11	36.4m	23.3

93

Summary of Statistical Data for Each Subject—*Continued*

Date	No. of Nights	Time in Bed		MRP1		Time Asleep		MRP2		Initial Active Period		
		Mean	σ	Mean	σ	Mean	σ	Mean	σ	No. of Nights	Mean	σ
Dec., 1929. . . .	10	11h 16m	68.0	8.01m	1.21	10h 42m	66.5	8.70m	1.39	14	31.1m	22.8
Both months. .	21	11h 32m	11.0	7.68m	1.09	10h 55m	53.0	8.32m	1.19	25	33.4m	23.2
						SUBJECT 14						
Nov., 1929. . . .	11	11h 12m	117.0	6.91m	1.63	9h 47m	96.5	7.79m	1.22	11	69.6m	93.8
Dec., 1929. . . .	24	10h 11m	53.0	7.64m	1.02	9h 55m	48.5	7.98m	1.04	25	9.8m	12.7
Jan., 1930. . . .	22	10h 46m	43.0	7.28m	0.76	10h 25m	34.5	7.60m	0.70	23	19.2m	18.1
All months. . .	57	10h 36m	69.5	7.36m	1.11	10h 05m	57.5	7.79m	1.03	59	14.6m	48.2
						SUBJECT 15						
April, 1929. . .	6	11h 03m	28.0	7.05m	0.59	10h 04m	32.5	8.02m	0.73	6	29.2m	11.0
May, 1929. . .	24	10h 40m	102.5	6.83m	0.84	9h 50m	43.0	7.67m	1.17	24	31.9m	16.2
June, 1929. . .	6	10h 47m	30.0	6.41m	0.29	9h 39m	60.0	7.40m	0.41	6	36.7m	9.4
Nov., 1929. . .	15	10h 58m	82.5	6.16m	0.51	10h 13m	43.0	6.76m	0.73	16	29.1m	12.1
Dec., 1929. . .	19	10h 54m	61.0	5.53m	0.53	9h 58m	59.5	6.08m	0.73	20	43.0m	18.5
All months. . .	70	10h 50m	80.5	6.32m	0.82	9h 57m	53.5	7.05m	1.14	72	34.5m	16.3
						SUBJECT 16						
April, 1929. . .	1	10h 55m	...	6.67m	...	9h 53m	...	7.58m	...	1	60.0m	...
May, 1929. . .	28	10h 45m	27.0	6.37m	0.73	9h 48m	33.5	7.15m	0.93	28	40.9m	22.4
June, 1929. . .	6	10h 44m	40.5	6.42m	0.62	9h 33m	20.5	7.35m	0.32	6	64.2m	21.5
All months. . .	35	10h 45m	29.5	6.39m	0.66	9h 45m	33.0	7.20m	0.82	35	49.5m	23.0
						SUBJECT 18						
April, 1929. . .	2	9h 53m	32.5	7.82m	0.74	9h 33m	15.0	8.35m	1.19	2	17.5m	17.5
May, 1929. . .	26	9h 52m	47.0	6.91m	0.93	9h 20m	46.0	7.44m	0.90	27	16.0m	14.5

June, 1929	7	9h 59m	31.5	6.99m	0.65	9h 17m	28.5	7.56m	0.67	7	17.2m	12.2
Nov., 1929	11	9h 54m	58.5	6.83m	0.57	9h 26m	54.5	7.25m	0.67	12	24.2m	21.4
Dec., 1929	7	11h 12m	25.5	6.61m	0.56	10h 20m	36.0	7.31m	0.62	9	16.1m	18.6
Jan., 1930	14	10h 57m	40.5	6.24m	1.34	10h 10m	31.0	6.73m	1.40	15	39.0m	27.1
All months	67	10h 15m	53.5	6.76m	0.98	9h 35m	49.0	7.29m	1.00	72	22.3m	21.3

SUBJECT 19

April, 1929	19	10h 27m	32.5	6.92m	0.82	9h 45m	44.5	7.61m	0.89	22	32.3m	11.4
May, 1929	10	10h 03m	65.0	6.87m	0.95	9h 10m	68.5	7.73m	0.87	10	39.0m	13.4
Both months	29	10h 19m	50.5	6.91m	0.81	9h 33m	59.0	7.65m	0.90	32	34.4m	12.4

SUBJECT 20

April, 1929	18	10h 24m	64.0	7.73m	0.78	9h 57m	55.0	8.21m	1.25	18	25.3m	14.2
May, 1929	15	9h 39m	47.5	7.02m	0.90	9h 16m	38.0	7.41m	0.98	15	20.4m	13.3
June, 1929	3	10h 32m	58.0	6.85m	0.34	9h 49m	41.0	7.49m	0.34	3	40.0m	18.7
Nov., 1929	1	11h 20m	...	7.43m	...	10h 28m	...	8.36m	...	1	50.0m	...
Dec., 1929	5	9h 48m	40.5	7.14m	0.69	9h 19m	50.0	7.68m	0.64	6	25.0m	16.9
All months	42	10h 06m	58.5	7.33m	0.92	9h 38m	54.0	7.84m	0.89	43	25.1m	15.8

SUBJECT 21

May, 1929	27	10h 57m	50.5	7.37m	0.91	10h 09m	47.0	8.21m	0.94	28	44.1m	24.2
June, 1929	7	10h 48m	32.5	6.52m	0.90	9h 48m	31.0	7.34m	1.02	7	57.9m	19.1
Nov., 1929	12	11h 34m	19.5	6.80m	0.85	11h 01m	33.0	7.28m	1.01	12	23.4m	11.9
Dec., 1929	20	11h 23m	16.5	6.98m	0.74	10h 45m	31.5	7.55m	0.85	21	23.6m	13.5
All months	66	11h 11m	40.5	7.06m	0.88	10h 27m	45.0	7.75m	1.00	68	35.5m	22.6

SUBJECT 32

April, 1929	9	10h 15m	51.0	10.32m	1.04	9h 47m	36.5	10.57m	5.04	9	25.6m	21.4
May, 1929	27	10h 33m	38.5	11.66m	2.34	10h 30m	37.5	13.09m	2.45	27	27.6m	14.5
June, 1929	7	10h 35m	38.0	9.35m	1.07	9h 49m	33.0	10.83m	2.17	7	37.9m	17.9
Nov., 1929	12	11h 19m	49.5	9.37m	1.45	10h 49m	51.0	10.06m	2.01	12	24.6m	16.6
Dec., 1929	19	11h 18m	27.5	8.64m	1.26	10h 35m	58.0	9.70m	1.44	19	40.8m	18.0
All months	74	10h 55m	48.0	10.13m	2.15	10h 16m	52.5	11.33m	2.51	74	31.2m	18.3

BIBLIOGRAPHY

1. ANONYMOUS. Book review of Paul Karger: Uber den Schlaf des Kindes. American Journal of Diseases of Children, 30: 601–02 (1925).
2. ANONYMOUS. The sleep of young children: report to the parents who participated in the study. (Institute of Child Welfare, University of Minnesota, Circular No. 4). 1930. 11 pp.
3. BOYNTON, M. ADELIA, and FLORENCE L. GOODENOUGH. The posture of nursery school children during sleep. American Journal of Psychology, 42: 270–78 (1930).
4. CZERNY, A. Beobachtungen über den Schlaf im Kindesalter unter physiologischen Verhältnissen. Jahrbuch für Kinderheilkunde, 33: 1–28 (1892).
5. DAVIS, C. The self-selected diet of a newly weaned infant. Journal of American Dental Association, 14: 1119 (1927).
6. ————. Self-selection of diet by newly weaned infants. American Journal of Diseases of Children, 36: 651 (1928).
7. FOSTER, JOSEPHINE C., FLORENCE L. GOODENOUGH, and JOHN E. ANDERSON. The sleep of young children. Journal of Genetic Psychology, 35: 201–17 (1928).
8. FOSTER, JOSEPHINE C., and MARION L. MATTSON. Nursery school procedure. D. Appleton, New York, 1929. Pp. 147–54.
9. GARVEY, C. R. An experimental study of the sleep of preschool children. (Abstract.) Proceedings of the Ninth International Congress of Psychologists, New Haven, 1929. Pp. 176–77.
10. ————. Sigmas of combined distributions calculated from sigmas, means, and frequencies of component distributions. Journal of Educational Psychology, 22: 307–10 (1931).
11. GERBER, WILHELM. Uber den Schlaf des Menschen und einen in der Praxis verwendbaren Schlafkontrollapparat. Münchener medizinische Wochenschrift, 69: 1399–400 (1922).
12. ————, and Gabriele Rembold. Untersuchungen über die Wirkung einzelner Schlafmittel (Dicodid, Codeonal, Paracodin und Bromural) mit dem Schlafkontrollapparat. Münchener medizinische Wochenschrift, 70: 1386–88 (1923).
13. GOODENOUGH, FLORENCE L. Inter-relationships in the behavior of young children. Child Development, 1: 29–47 (1930).
14. HAAS, ALBERT. Uber Schlaftiefenmessungen. Psychologische Arbeiten, 8: 228 (1923).
15. HARRIS, J. ARTHUR. The arithmetic of the product-moment method of calculating the coefficient of correlation. American Naturalist, 44: 693–99 (1910).
16. HAYASHI, Y. On the sleeping hours of school children of 6 to 20 years. Jido Jatshi (Child's Journal), 1925. Abstracted by J. G. Yoshioka in Psychological Abstracts, No. 1826, 1927.
17. HERZ, FRANZ. Selbstbeobachtung über freiwillige Schlafentziehung. Pflügers Archiv für die gesamte Physiologie, 200: 429–42 (1923).
18. HOLZINGER, KARL J. Statistical tables for students in education and psychology. University of Chicago Press, 1925.
19. HOWELL, W. H. A contribution to the physiology of sleep, based on plethysmographic experiments. Journal of Experimental Medicine, 2: 313–46 (1897).

20. HUNTER, W. S. Foundations of experimental psychology. Clark University Press, Worcester, Massachusetts, 1929. Chapter 15.
21 JOHNSON, H. M. An essay toward an adequate explanation of sleep. Psychological Bulletin, 23: 141–42 (1926).
22. ————. How we sleep and why. Columbia, Vol 8, No. 9, pp. 34–35, 39, 41 (1929).
23. ————. Is sleep a vicious habit? Harper's, 157: 731–41 (1928).
24. ————. The measurement of sleep. Hospital Progress, 8:351–63 (1927).
25. ————. Some further experiments bearing on the problem of sleep. (Abstract.) Psychological Bulletin, 24: 516–17 (1927).
26. ————, and T. H. SWAN. Sleep. Psychological Bulletin, 27: 1–39 (1930).
27. ————, T. H. SWAN, and G. E. WEIGAND. Sleep. Psychological Bulletin, 23: 482–503 (1926).
28. ————, and G. E. WEIGAND. The measurement of sleep. Proceedings of Pennsylvania Academy of Science, 2: 43–48 (1927).
29. KARGER, P. Uber den Schlaf des Kindes. Karger, Berlin, 1925. iv + 50 pp.
30. ————. Uber den Schlaf und die Schlafbewegungen des Kindes. Beihefte zu Jahrbuch für Kinderheilkunde, Heft 4 (1925).
31. ————. Unsere heutigen Kentnisse über den Schlaf, Fortschritte der Medizin, 42: 237–39 (1924).
32. KELLEY, TRUMAN L. Statistical method. Macmillan Co., New York, 1924.
33. KJERSTAD, C. L. The form of learning curves for memory. Psychological Monographs, Vol. 26, No. 5 (1919).
34. KLEITMAN, N. The effects of prolonged sleeplessness on man. American Journal of Physiology, 66: 67–92 (1923).
35. KOHLSCHÜTTER, E. Messungen der Festigkeit des Schlafes. Zeitschrift für rationelle Medizin, Series 3, 17: 209–53 (1863).
36. KREIDL, A., and F. HERZ. Der Schlaf des Menschen bei Fernbleiben von Gesichts- und Gebörseindrücken. (Uber den Schlaf der Mindersinnigen.) Pflügers Archiv für die gesamte Physiologie, 203: 459–71 (1924).
37. LANDIS, C. Electrical phenomena of the body during sleep. American Journal of Physiology, 81: 6–19 (1927).
38. LASLETT, H. A. An experiment on the effects of loss of sleep. Journal of Experimental Psychology, 7: 45–58 (1924).
39. LEE, MARY A. M., and N. KLEITMAN. Studies in the physiology of sleep. II. Attempts to demonstrate functional changes in the nervous system during experimental insomnia. American Journal of Physiology, 67: 141–52 (1923).
40. MICHELSON, F. Untersuchungen über die Tiefe des Schlafes. Schnakenburgs Buchdruckerie, Dorpat, 1891.
41. MILES, W. R. Duration of sleep and the insensible perspiration. Proceedings of the Society for Experimental Biology and Medicine, 26: 577–80 (1929).
42. MÖNNINGHOFF, O., and F. PIESBERGEN. Messungen über die Tiefe des Schlafes. Zeitschrift für Biologie, 19: 114–28 (1883).
43. PATRICK, G. T. W., and J. A. GILBERT. On the effects of loss of sleep. Psychological Review, 3: 469–83 (1896). Reprinted in University of Iowa Studies in Psychology, No. 1 (1897).
44. PEIPER, A. Untersuchungen über den galvanischen Reflex. Jahrbuch für Kinderheilkunde, 107: 139 (1924).
45. RENSHAW, S., and A. P. WEISS. Apparatus for measuring changes in bodily posture. American Journal of Psychology, 37: 261–27 (1926).
46. ROBINSON, E. S., and W. T. HERON. Result of variations in length of memorized material. Journal of Experimental Psychology, 5: 428–48 (1922).
47. ————, and S. O. HERRMANN. Effects of loss of sleep. I. Journal of Experimental Psychology, 5: 19–24 (1922).

48. ————, and F. RICHARDSON-ROBINSON. Effects of loss of sleep. II. Journal of Experimental Psychology, 5: 93–100 (1922).

49. SANCTIS, S. DE, and U. NEYROZ. Experimental investigations concerning the depth of sleep. (Translated by H. C. Warren.) Psychological Review, 9: 254–82 (1902).

50. SHIRLEY, MARY. Spontaneous activity. Psychological Bulletin, 26: 341–65 (1929).

51. SIDIS, BORIS. An experimental study of sleep. Journal of Abnormal Psychology, 3: 1 (1908).

52. STARLING, E. H. Principles of human physiology. 3d ed., Lea and Febiger, Philadelphia, 1920.

53. STEWART, C. C. Variations in daily activity produced by alcohol, by changes in erometric pressure and diet, with a description of recording methods. American Journal of Physiology, 1: 39–56 (1898).

54. SWAN, T. H. A note on Kohlschütter's curve of the "depth of sleep." Psychological Bulletin, 26: 607–10 (1929).

55. SZYMANSKI, J. S. Aktivität und Ruhe bei der Menschen. Zeitschrift für angewandte Psychologie, 20: 192–222 (1922).

56. ————. Aktivität und Ruhe bei Tieren und Menschen. Zeitschrift für allgemeine Physiologie, 18: 105–62 (1920).

57. ————. Eine Methode zur Untersuchung der Ruhe- und Aktivitätsperioden bei Tieren. Pflügers Archiv für die gesamte Physiologie, 158: 343–85 (1914).

58. ————. Versuche über die Aktivität und Ruhe bei Säuglingen. Pflügers Archiv für die gesamte Physiologie, 172: 424–29 (1918).

59. ————. Die Verteilung von Ruhe- und Aktivitätsperioden bei einigen Tierarten. Pflügers Archiv für die gesamte Physiologie, 158: 343–85 (1914).

60. VINCENT, STELLA B. The function of the vibrissae in the behavior of the white rat (Animal Behavior Monographs, Vol. I, No. 5). Williams and Wilkins, Baltimore, 1912.

61. WARREN, RICHARD, and ROBERT M. MENDENHALL. The Mendenhall-Warren-Hollerith correlation method (Statistical Bureau Document No. 1). Columbia University, 1929.

INDEX

Made in the USA
Monee, IL
07 July 2026

56552238R00066